Silicon and Plant Diseases

Fabrício A. Rodrigues • Lawrence E. Datnoff
Editors

Silicon and Plant Diseases

Springer

Editors
Fabrício A. Rodrigues
Departamento de Fitopatologia
Universidade Federal de Viçosa
Viçosa, Minas Gerais, Brazil

Lawrence E. Datnoff
Department of Plant Pathology
and Crop Physiology
Louisiana State University
Agricultural Center
Baton Rouge, LA, USA

ISBN 978-3-319-22929-4 ISBN 978-3-319-22930-0 (eBook)
DOI 10.1007/978-3-319-22930-0

Library of Congress Control Number: 2015951978

Springer Cham Heidelberg New York Dordrecht London

Printed on acid-free paper

Springer International Publishing AG Switzerland is part of Springer Science+Business Media (www.springer.com)

Foreword

Silicon (Si) is the most abundant mineral element in the earth's crust, so all plants rooting in soil contain Si in quantities that exceed many essential mineral elements. However, due to its universal existence and lack of obvious visible deficiency symptoms, very little attention was paid to the role of this element in plant growth until 1917, when the first scientific report on Si suppressing a plant disease was published by Onodera in Japanese. Onodera found that rice leaves with a low Si content was susceptible to blast while leaves with high Si were more resistant to this disease. Nowadays, Si is known to suppress many plant diseases in both monocots and dicots such as bacterial blight, brown spot, grain discoloration, leaf scald, leaf and panicle blast, stem rot, and sheath blight in rice as well as powdery mildew in wheat and cucumber. Mechanisms underlying the Si-mediated resistance to different diseases have also been intensively studied.

In this book written by Drs. Rodrigues from Brazil and Datnoff from USA, the authors comprehensively cover all aspects on the relationship between Si and plant disease that span from history to disease control, mechanisms involved and finally to future prospects. Both authors are excellent scientists, who have been working in the field of Si and disease control together for many years. This book widely collects data from their own research and other groups around the world.

Yield loss due to disease is a major problem in crop production worldwide, therefore control of disease occurrence is an important issue. Different from most fungicides, Si is able to decrease the intensity of multiple diseases at the same time. In some cases, Si fertilizers have been demonstrated to be as effective as fungicides in reducing pathogen infection. Therefore, application of Si fertilizers has become a routine agronomic practice for sustainable crop production. In fact, Si has been recognized as an "agronomically essential element" for rice production in Japan, and Si fertilizers have been applied in many countries such as Brazil, China, Japan, and the USA. Since the effect of Si is more obvious in reducing disease intensity under intensive cultivation with heavy applications of nitrogen fertilizers, the demands for Si fertilizers will increase in the future. This book will provide very useful information on how Si controls plant diseases not only for students at the university, but also for researchers in other agricultural fields of study.

Professor, Institute of Plant Science and Resources, Jian Feng Ma
Okayama University, Japan

Preface

Silicon, considered to be the second-most abundant mineral element in soil, plays an important role in the mineral nutrition of plants. A wide variety of monocot and dicot species have benefited from silicon nutrition, whether direct or indirect, when they are exposed to different types of abiotic and/or biotic stresses. Besides the many agronomic and horticultural benefits gained by maintaining adequate levels of this element in the soil and also in the plant tissue, the most notable effect of silicon is the reduction in the intensities of a number of plant diseases caused by biotrophic, hemibiotrophic and necrotrophic plant pathogens in many crops of great economic importance.

The aim of this book is to summarize our current understanding of the effects of silicon on plant diseases. The chapters address the dynamics of silicon in soils and plants; the history of silicon in the control of plant diseases; the use of silicon to control soil-borne, seed-borne and foliar diseases in monocots and dicots; the mechanisms involved in the host resistance against infection by plant pathogens mediated by silicon as well as the current knowledge at the omics level and, finally, highlights and prospects for using silicon in the future. We hope this book will be a valuable asset for managing plant diseases as well as a useful resource for undergraduate and graduate courses in plant pathology and other related disciplines. We believe the in-depth information found in this book will be useful to plant scientists worldwide and of interest to agronomists, horticulturists, plant pathologists, plant physiologists and soil scientists in its references to numerous commodity crops, ornamentals and turf. As researchers and growers become more aware of silicon and its potential, it is likely that this often overlooked, quasi-essential element will be recognized as a viable means of enhancing plant health and performance.

Viçosa, Minas Gerais, Brazil — Fabrício A. Rodrigues
Baton Rouge, LA, USA — Lawrence E. Datnoff

Acknowledgments

Prof. Fabrício A. Rodrigues thanks the National Council for Scientific and Technological Development (CNPq), the Coordination for the Improvement of Higher Education Personnel (CAPES), and The Minas Gerais State Research Foundation (FAPEMIG) for the grants that supported the studies regarding the use of silicon for the control of plant diseases in the Laboratory of Host-Parasite Interaction of the Plant Pathology Department at Viçosa Federal University. Prof. Fabrício A. Rodrigues also thanks the CNPq for his fellowship. Prof. Lawrence E. Datnoff thanks the Louisiana State University Agricultural Center for their generous operational and financial support as well as past funding from USAID, USDA-OICD, USDA-FAS and the Conservation, Food and Health Foundation, Inc. The authors thank all their undergraduate and graduate students for their interest, dedication and efforts in conducting the numerous experiments involved in the use of silicon to control plant diseases. We also gratefully acknowledge the support and assistance of the publishers Elsevier Science Publishers (Crop Protection) and the American Phytopathological Society (Phytopathology) and the Brazilian Society of Plant Pathology (Tropical Plant Pathology) for the use of their pertinent data and information. All other references cited in this book are also acknowledged with gratitude.

Contents

Contributors

Leandro José Dallagnol Departamento de Fitossanidade, Universidade Federal de Pelotas, Pelotas, Rio Grande do Sul, Brazil

Lawrence E. Datnoff Department of Plant Pathology and Crop Physiology, Louisiana State University Agricultural Center, Baton Rouge, LA, USA

Henrique Silva Silveira Duarte Departamento de Proteção de Plantas, SCA-DFF, Universidade Federal do Paraná, Curitiba, Paraná, Brazil

Alessandro Antônio Fortunato Departamento de Fitopatologia, Universidade Federal de Viçosa, Viçosa, Minas Gerais, Brazil

Joseph Raymond Heckman Department of Plant Biology and Pathology, Rutgers University, New Brunswick, NJ, USA

Renata Sousa Resende Departamento de Fitopatologia, Universidade Federal de Viçosa, Viçosa, Minas Gerais, Brazil

Fabrício A. Rodrigues Departamento de Fitopatologia, Universidade Federal de Viçosa, Viçosa, Minas Gerais, Brazil

Brenda Servaz Tubaña School of Plant, Environmental, and Soil Sciences, Louisiana State University Agricultural Center, Baton Rouge, LA, USA

Chapter 1
History of Silicon and Plant Disease

Lawrence E. Datnoff and Fabrício A. Rodrigues

Abstract The use of silicon in agriculture probably began in China more than 2000 years ago because farmers at that time incorporated rice straw along with manure as a fertilizer to enhance plant performance and yield. In 1917, the potential of silicon to reduce blast on rice was first reported by a plant chemist, and his discovery launched a cascade of silicon research in Japan. The role of silicon in plant growth and potential disease reduction was first noted for dicots in 1939. As a result of research from the 1980s until today, silicon's potential to decrease the intensity of many diseases is now known for a large number of plant species. Since the early discovery that this quasi-essential element believed to be unimportant in plant development plays a major role in reducing plant diseases, research has revealed that amending plants with silicon is a simple, sustainable way to help maintain and enhance plant health in agriculture.

Introduction

The use of silicon in agriculture most likely began more than 2000 years ago in China (Matichenkov et al. 2001). At that time, the emperor decreed that farmers must incorporate rice straw along with manure as a fertilizer to enhance plant performance and yield (Yoshida 1978). Because rice plant tissue is now known to contain anywhere from 1 to 10 dag/kg silicon, this would be the first indirect evidence of early agriculturists using silicon as a soil amendment/fertilizer.

In the early 1800s, plant naturalists began to measure the elemental composition of a number of plant species and discovered that plants contained silicon in quantities that greatly exceeded those of other mineral elements (De Saussure 1804;

L.E. Datnoff (✉)
Department of Plant Pathology and Crop Physiology, Louisiana State University Agricultural Center, 302 Life Sciences, Baton Rouge, LA 70803, USA
e-mail: LDatnoff@agcenter.lsu.edu

F.A. Rodrigues
Departamento de Fitopatologia, Universidade Federal de Viçosa, Viçosa, Minas Gerais Cep 36570-900, Brazil
e-mail: fabricio.plantpathologist@gmail.com; fabricio@ufv.br

F.A. Rodrigues, L.E. Datnoff (eds.), *Silicon and Plant Diseases*,
DOI 10.1007/978-3-319-22930-0_1

Snyder et al. 2006; Tottingham 1908). Even with these findings, the general scientific community at that time considered silicon to be unessential for plant development (Tottingham 1908). Davy (1819), however, was one of the first scientists to investigate the form of silicon in the epidermis of a number of monocots, including horsetail, oats and wheat. This author believed that the use of silicon in agriculture might be important because the plant's epidermis was siliceous and probably served to support, as well as protect, the plant from biotic stresses. Liebig (1840) surmised that silicon was involved in cereal stalk rigidity and that the lodging of wheat was simply caused by a deficiency in this element. Leibig further suggested that sodium silicate could be used as a silicon fertilizer and conducted research with this silicon source on sugar beet development under greenhouse conditions. In 1842, Berzelius discovered silicon as an element and studied its role with organic matter under field conditions (Matichenkov et al. 2001). Kreuzhage and Wolff (1884) and Grob (1896) first studied microscopically the distribution and specific location of silicon bodies in different leaf tissue of oats. Based on these authors' observations that the cell lumen contained a high level of silicon, Grob (1896) suggested that this element might be involved in resistance against plant diseases.

Discovery of Silicon Affecting Plant Diseases in Monocots

Isenosuke Onodera in 1917 was probably the first scientist to demonstrate that silicon may play a role in reducing plant disease (Onodera 1917; Ishiguro 2001). As such, the history of this element in association with a known plant disease infecting a monocot began with his studies. This scientist was a plant chemist by training and was interested in determining whether rice infected by *Pyricularia oryzae*, the causal agent of rice blast, would reduce the mineral content of the plant. He collected rice plants that were visibly non-infected and infected with *P. oryzae* in the same field from 13 Western provinces of Japan. He then compared the nutritional compositions of these plants to each other and observed that the plants infected with *P. oryzae* contained a lower silicon content than those not infected. Therefore, Onodera demonstrated a potential relationship between the silicon content and rice susceptibility to blast. Furthermore, he deduced that the silicon content of the rice plant was probably dependent on the soil type in which the plant was grown, suggesting that some soils are inherently lower in elemental silicon compared to others.

Onodera's discovery launched a cascade of silicon research on rice blast in Japan (Ishiguro 2001; Suzuki 1963; Kozaka 1963). Miyake and Adachi (1922) demonstrated that rice cultivars that contained more silicon were more resistant to blast than those with lower concentrations of this element. Kawashima (1927) showed that increasing silicon rates in the soil in which rice plants were grown increased the concentration of silicon in the plant tissue and subsequently decreased blast development. Other investigators in the 1930s also showed that the application of silicon increased rice resistance to blast (Ishiguro 2001). Over the next 20 years, researchers focused on the mechanism of silicon-mediated host resistance to diseases – whether it might be mechanical or biochemical. Regarding the mechanical hypothesis, a number of investigators studied the relationship between silicon rates and the number of silicified cells that formed in the epidermis, while the focus of biochemical studies was primarily on puncture resistance, i.e., the higher silicon content in the plant lowered the ability of *P. oryzae* to penetrate the rice epidermal cells and cause infection. From these studies, Yoshida and his colleagues (1962) believed that a 2.5-μm-thick layer of silicon was deposited just below the space beneath the thin cuticle, and the mechanical barrier hypothesis (silicon-cuticle double layer) was subsequently proposed to explain how silicon conferred rice resistance to blast. In the 1950s, Suzuki and Shigematsu introduced the use of calcium silicate slag as a source of soluble silicon for controlling rice blast while improving yields at rates ranging from 0.2 to 16 t of product per ha; this application has become a common agricultural practice for rice production in Japan (Ishiguro 2001).

While the Japanese were studying the effects of silicon against rice blast, researchers in Germany and the US were discovering its effect against other plant diseases in millet and wheat (Sommer 1926; Germar 1934). Sommer (1926) demonstrated that silicon greatly enhanced the seed yield of millet and that the seed heads of those plants grown without silicon were severely infected by plant pathogenic fungi. Germar (1934) reported a reduction in the development of powdery mildew on wheat with silicon and suggested that this enhanced resistance was due to an increase in silicified leaf tissue. This author also observed that silicon was not as effective if the fungus entered through the stomata.

Discovery of Silicon Affecting Plant Diseases in Dicots

Raleigh (1939) investigated the role of silicon in the growth of table beets and may have been the first scientist to suggest that silicon affects plant disease in dicots. This author observed that beet plants had greater shoot and root mass when grown in nutrient culture with silicon compared to those grown without the element. He also observed that plant roots became necrotic and were covered by an organismal growth, probably a fungal or fungal-like mass, when grown in a nutrient solution without silicon. Upon further investigation, this author found that plants transferred to a silicon-deficient solution at an early stage of growth suffered from damping off. However, when silicon was added, practically no damping off occurred. Based on these findings, he concluded that silicon played an important role in the growth and development of table beets.

Wagner (1940) was interested in determining whether this element could suppress powdery mildew development on cucumbers, especially because an early study had demonstrated that silicon was effective in suppressing powdery mildew development in wheat (Germar 1934). This author was possibly the first scientist to demonstrate through control experiments that silicon could suppress infection by an obligate biotrophic fungus in a dicot. He used sodium silicate and calcium silicate as the silicon sources, which were compared to a non-treated control. He found that sodium silicate and calcium silicate dramatically reduced the number of conidial colonies and necrotic spots by 55 % and 73 %, respectively, compared to the control treatment. As observed by Germar (1934) for powdery mildew on wheat, Wagner believed that silicon impeded the development of the haustoria formed by the fungus on the leaves of cucumbers.

Conclusions

By the 1950s, researchers in China and Korea began to investigate the effects of silicon on rice growth and yields and found that this element could significantly decrease blast development as well as the intensity of other important diseases of rice (Wang et al. 2001; Park 2001). During this same time period, the beneficial effects of silicate materials were first demonstrated for sugarcane growth and yield in Hawaii (Plucknett 1971). Researchers also discovered that silicon could alleviate a physiological disease in sugarcane known as 'freckling', in which tiny clear spots on the leaf blade gradually become necrotic and then coalesce to the point where no healthy green plant tissue is apparent. To date, freckling is the only known nutritional plant disorder ever described that mimics the symptoms of a plant disease and could be attributed to a silicon deficiency.

As a result of research conducted from the 1980s until today, silicon is now known to dramatically decrease the intensity of diseases, such as damping off, leaf blights, leaf spots, galls, powdery mildews, root rots, rusts and wilts, of many important plant species. These diseases are caused by all the major plant pathogen groups, including approximately 83 that are fungal, nine fungal-like (Oomycota), six bacterial, one viral and three nematodal, with the nematode belonging to the genus *Meloidogyne*. Since the early discovery that this quasi-essential element believed to be unimportant in plant development plays a major role in reducing plant diseases, research has revealed that amending plants with silicon is a simple, sustainable way to help maintain and enhance plant health in agriculture (Datnoff et al. 2007).

References

Datnoff LE, Rodrigues FA, Seebold K (2007) Silicon and plant disease. In: Datnoff LE, Elmer WH, Huber DM (eds) Mineral nutrition and plant disease. The American Phytopathological Society, St. Paul, pp 233–246

Davy H (1819) Elements of agricultural chemistry: in a course of lectures for the Board of Agriculture. Hudson and Company Printers, Hartford

De Saussure NT (1804) Recherches chimiques sur la végétation. Chez la veuve Nyon, Paris

Germar B (1934) Über einige Wirkungen der Kieselsäure in Getreidepflanzen, insbesondere auf deren Resistenz gegenüber Mehltau. Zeitschriftfür für Pflanzenernährung 35:102–115

Grob A (1896) Beitrage zur anatomie der epidermis der gramineen blatter. Biblioth Bot 7:1–64
Ishiguro K (2001) Review of research in Japan on the roles of silicon in conferring resistance against rice blast. In: Datnoff LE, Snyder GH, Korndörfer GH (eds) Silicon in agriculture, Studies in plant science, vol 8. Elsevier Science, Amsterdam, pp 277–291
Kawashima R (1927) Influence of silica on rice blast disease. Jpn J Soil Sci and Plant Nutr 1:86–91
Kozaka T (1963) Control of rice blast by cultivation practices in Japan. In: The rice blast disease, International Rice Research Institute. The John Hopkins Press, Baltimore, pp 421–438
Kreuzhage C, Wolff E (1884) Bedeutung der Kieselsaure fur die Entwicklung der Haferpflanze Versuchen in Wasserkultur. Organ F Naturw Forschugen Landw Vers Stat Dresden 30:169–197
Liebig JF (1840) Organic chemistry and its applications to agriculture and physiology. Taylor and Walton, London
Matichenkov VV, Bocharnikova EA, Datnoff LE (2001) A proposed history of silicon fertilization. In: Datnoff LE, Snyder GH, Korndörfer GH (eds) Silicon in agriculture, Studies in plant science, vol 8. Elsevier Science, Amsterdam, p 36
Miyake Y, Adachi M (1922) Chemische untersuchungen uber die Widerstandsfahigkeit der Reisarten gegen die Inokrankheit 1–2 bericht. J Biochem Japan 1:223–247
Onodera I (1917) Chemical studies on rice blast. J Sci Agric Soc 180:606–617
Park CS (2001) Past and future advances in silicon research in the Republic of Korea. In: Datnoff LE, Snyder GH, Korndörfer GH (eds) Silicon in agriculture, Studies in plant science, vol 8. Elsevier Science, Amsterdam, pp 359–371
Plucknett DL (1971) The use of soluble silicates in Hawaiian agriculture. Univ Queensland Publ 1:203–223
Raleigh GJ (1939) Evidence for the essentiality of silicon for growth of the beet plant. Plant Physiol 14:823–828
Snyder GH, Matichenkov VV, Datnoff LE (2006) Silicon. In: Barker A, Pilbeam D (eds) Handbook of plant nutrition. Taylor and Frances, Boca Raton, pp 551–568
Sommer AL (1926) Studies concerning the essential nature of aluminum and silicon for plant growth. Univ Calif Publ Agric Sci 5:57–81
Suzuki N (1963) Nature of resistance to blast. In: The rice blast disease, International Rice Research Institute. The John Hopkins Press, Baltimore, pp 277–301
Tottingham WE (1908) The status of silicon in certain plants. Master of Science, University of Wisconsin, Madison
Wagner F (1940) Die Bedeutung der Kieselsäure für das Wachstum einiger Kulturpflanzen, ihren Nahrstoffhaushalt und ihre Anfälligkeit gegen echte Mehltaupilze. Phytopathol Ztschr 12:427–479
Wang H, Chunhua L, Liang Y (2001) Agricultural utilization of silicon in China. In: Datnoff LE, Snyder GH, Korndörfer GH (eds) Silicon in agriculture, Studies in plant science, vol 8. Elsevier Science, Amsterdam, pp 343–358
Yoshida S (1978) The availability of silicon in paddy soil. In: Kawagushi K (ed) Paddy soil science. Kodansha, Tokyo, pp 293–299
Yoshida S, Ohnishi Y, Kitagishi K (1962) Histochemistry of Si in rice tissues. III. The presence of cuticle-silica double layer in the epidermal tissue. Soil Sci Plant Nutr 8:1–5

Chapter 2
Silicon in Soils and Plants

Brenda Servaz Tubaña and Joseph Raymond Heckman

Abstract The crust of the earth is largely composed of silicon that is found primarily as silicate minerals, secondary alumino silicates and various forms of silicon dioxide. However, the abundance of silicon in soils is not an indication that sufficient supplies of soluble silicon are available for plant uptake. In this chapter, the outcomes of many years of research conducted on silicon are consolidated to understand the state of knowledge for silicon fertilization guidelines in crop production. Monosilicic acid (H_4SiO_4) is the form of silicon used by plants, which is found both in liquid and adsorbed phases of silicon in soils. The concentration of the H_4SiO_4 in the soil solution is influenced by the soil pH and the amounts of clay, minerals, organic matter and Fe/Al oxides/hydroxides, which are collectively related to the geologic age of the soil. Fertilization can rapidly increase the concentration of H_4SiO_4 in the soil solution; therefore, fertilization has become a common practice in areas with intensive cropping systems, particularly for those soils that are inherently low in soluble silicon. The establishment of procedures to estimate the plant-available silicon and the critical soil silicon levels and the method (5-day Na_2CO_3-NH_4NO_3 extraction) to analyze the soluble silicon fraction in solid fertilizers were among the advances in research on silicon in agriculture in recent years. These measurements were the key components required for the development and implementation of effective silicon fertilizer management in crop production. However, many aspects of the role of silicon in soil science remain understudied, and these aspects should be the focus of future research.

B.S. Tubaña (✉)
School of Plant, Environmental, and Soil Sciences, Louisiana State University Agricultural Center, 104 Sturgis Hall, Baton Rouge, LA 70803, USA
e-mail: btubana@agcenter.lsu.edu

J.R. Heckman
Department of Plant Biology and Pathology, Rutgers University,
Foran Hall/Cook Campus, 59 Dudley Road, New Brunswick, NJ 08901-8520, USA

F.A. Rodrigues, L.E. Datnoff (eds.), *Silicon and Plant Diseases*,
DOI 10.1007/978-3-319-22930-0_2

Introduction

The silicon in the crust of the earth is ubiquitous. Over 100 years ago, man linked silicon fundamentally to his life style with the use of silicon in the household, in industrial applications, and in construction (Vasanthi et al. 2012). In agriculture, silicon is a nutrient for which an enormous amount of literature examines the value of silicon fertilization in improving overall crop productivity and health. Silicon is abundant in the soil but is primarily in an inert form and consequently is unavailable for plant uptake. Although years of research have focused on understanding the role of silicon in plant growth and development, to date, this element has been determined to be essential only for scouring rushes and diatoms and other members of the yellow-brown or golden algae (Epstein 1999). Nevertheless, the benefits of silicon fertilization to crop production are too significant to be overlooked, and in some agricultural areas, fertilization with silicon is the common agronomic practice. Thus, a renewed research effort was directed to develop guidelines and management practices for silicon fertilization for a number of agronomic and horticultural crops. In this chapter, the results from many years of research on silicon in soils and plants were consolidated, and the analysis included the chemical dynamics of the different forms of silicon in soils, specifically the form used by plants, monosilicic acid (H_4SiO_4), the assimilation and the role of silicon in plants, the critical levels of silicon in soils and plants, the procedures to estimate the plant-available silicon in soils, and the potential sources of silicon.

Silicon in Soils

Silicon is the second most abundant element in the crust of the earth after oxygen, with a mean content of 28.8 % (weight) and an occurrence that ranges from 0.52 to ~47 wt% (McKeague and Cline 1963; Wedepohl 1995). In rocks, the concentrations of silicon range from 23 % (e.g., basalt) to 46.5 % (e.g., orthoquartzite) (Monger and Kelly 2002). Trace amounts of silicon are also in carbonaceous rocks such as the limestones and the carbonites (Monger and Kelly 2002). The silcretes are the component of derived soils that contain significant amounts of silicon (as high as 46 %). The amount of silicon in the petrocalcic horizon is much lower than (~8 %) that in the silcretes, and the amount of silicon in the minerals found in some highly weathered Oxisols such as bauxites and ferricretes is even less (Monger and Kelly 2002). Whereas most soils are abundant in silicon, certain soils contain low levels of this element, particularly the plant-available form of silicon. These soils include the Oxisols and the Ultisols, which are typically characterized as highly weathered, leached, acidic and low in base saturation (Foy 1992), and the Histosols, which contain high levels of organic matter and very low mineral contents (Snyder et al. 1986). Additionally, the soils that are composed of a large fraction of quartz sand

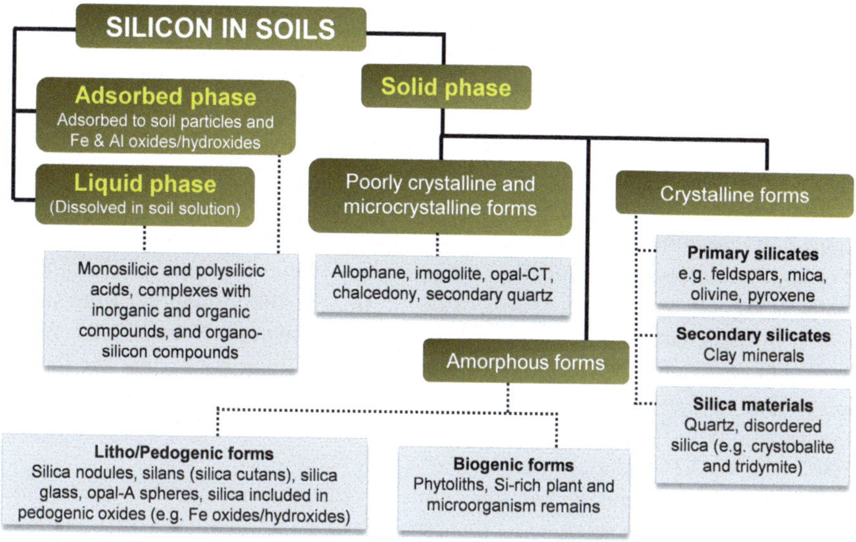

Fig. 2.1 Different fractions of silicon in soils (Modified from Matichencov and Bocharnikova (2001) and Sauer et al. (2006))

and those that have been under long-term crop production typically have low plant-available silicon (Datnoff et al. 1997).

In soils, silicon is generally grouped into three different fractions: the liquid phase, the adsorbed phase and the solid phase (Matichencov and Bocharnikova 2001; Sauer et al. 2006). The compositions of these different fractions are detailed in the classification of silicon compounds in soils that is presented in Fig. 2.1. The silica material was included by Sauer et al. (2006) among the crystalline forms of silicon in the solid phase fraction. Previously, the crystalline form consisted only of the primary and the secondary crystalline silicates, which are abundant in mineral soils that developed from rocks and sediments (Iler 1979; Conley et al. 2006). The silica materials consist primarily of quartz and disordered silica. The amorphous and poorly crystalline and microcrystalline forms are also components of the silicon fractions in the solid phase (McKeague and Cline 1963). The components of silicon in the liquid and the adsorbed phases are similar, with exception that those in liquid phase are dissolved in the soil solution, whereas those that are adsorbed are held onto soil particles and the Fe and Al oxides/hydroxides.

1. Solid Phase Silicon forms in the solid phase are divided into three primary groups: the amorphous forms, the poorly crystalline and microcrystalline forms, and the crystalline forms (Fig. 2.1). The largest fraction of silicon in the solid phase is the crystalline forms that occur primarily as primary and secondary silicates and silica materials. The primary mineral-bearing silicates that are inherited in soils are concentrated in the sand and silt particles; whereas the clay particles that are produced from the pedogenic processes that involve phylosilicates and Al-Fe oxides/

hydroxides contain the secondary silicates (Allen and Hajek 1989). Furthermore, Allen and Hajek (1989) found that silicon is also in poorly crystalline forms such as the short-range ordered silicates and in the microcrystalline forms such as chalcedony and secondary quartz. The formation of the short-range ordered silicates (e.g., allophane and imogolite) in soil horizons is favored with $pH_{H_2O}>5.0$ (Wada 1989), and the formation of the imogolite originates from the precipitation of H_4SiO_4 with Al hydroxides (Exley 1998; Doucet et al. 2001). Conversely, the formation of allophane and imogolite is inhibited in environments in which the decomposition rate of organic materials is high and the accumulation of humus is prevalent. The organic matter components bind the Al hydroxides to prevent the formation of the short-ranged silicates, but the formation of opaline silica is favored (Huang 1991). When the H_4SiO_4 concentration exceeds the solubility of the amorphous silica, the formation of opal-A, opal-CT and microquartz is promoted, whereas the secondary microcrystalline quartz is produced from the re-precipitation of the opal-CT from the dissolved opal-A (Chadwick et al. 1987).

The amorphous forms include the forms of both biogenic and litho/pedogenic origins and are in soils in amounts that range from <1 to 30 mg g^{-1} on a total soil basis (Jones 1969; Drees et al. 1989). The biogenic forms originate from plant residues and the remains of microorganisms and are collectively known as biogenic opal. The silicon absorbed by plants accumulates in the leaf, culm and stem as silica bodies or phytoliths, whereas the contributions of microorganisms are found as microbial and protozoic silicon (Sauer et al. 2006; Sommer et al. 2006; Aoki et al. 2007). The litho/pedogenic forms consist of silicon complexes with Al, Fe, heavy metals and soil organic matter (Matichencov and Bocharnikova 2001; Farmer et al. 2005). Furthermore, the pedogenic forms are characterized as the noncrystalline inorganic fractions, which include opal A, glasses and opal coatings on secondary minerals (McKeague and Cline 1963; Chadwick et al. 1987; Drees et al. 1989). The opal A is formed when the soluble silicon in the soil is at supersaturated levels (Drees et al. 1989). The formation of the opal coatings, such as the silcretes and the cements, is common in most soils and is classified as a secondary product of weathering (Dove 1995; Basile-Doelsch et al. 2005). According to Drees et al. (1989), the biogenic-based opal is commonly found in significant amounts under a wide range of environmental conditions, whereas the formation of the pedogenic opal occurs under specific physico-chemical soil conditions.

The solubility of the different forms of silicon in the solid phase significantly affects the concentration of silicon in the soil solution. The packing density of the silica tetrahedral and the long-range crystal order influences the solubility of the silica-bearing minerals (Iler 1979; Lindsay 1979; Drees et al. 1989). For example, larger contributions are expected from amorphous silica because of the higher solubility (1.8–2 mM silicon) than those from quartz. The dissolution rates of amorphous silica linearly increased with saturation but exhibited an exponential dependence on the electrolytes that was similar to quartz (Dove et al. 2008). The solubility of quartz ranged only from 0.10 to 0.25 mM silicon because quartz is highly stable and thermodynamically resistant to weathering (Drees et al. 1989; Monger and Kelly 2002). Thus, if quartz is ubiquitous in both the residual and the transported parent materials, the contribution of quartz to the silicon in soil solution will be minimal. Fraysse et al. (2006) noted that the solubility of biogenic-based silica is 17-fold higher than that of quartz. The silica contained in the phytoliths is classified as a pure inorganic pool, because the rate of release from plant litter is independent of cellulose hydrolysis, and the released silica does not form complexes with organic matter (Fraysse et al. 2010). The solubility of both the crystalline and the amorphous silica is approximately constant between pH values 2 and 8.5 but increases rapidly at pH ~9 because of the reduction of the H_4SiO_4 concentration in the soil solution. At pH ~9, the H_4SiO_4 dissociates to $H_3SiO_4^- + H^+$ (Dove 1995), which initiates the dissolution of the crystalline and the amorphous silica to replenish or buffer the reduced concentration of the H_4SiO_4 in the soil solution.

2. Silicon in Soil Solution Silicon is in the soil solution in different forms and occurs primarily as monomeric (H_4SiO_4, the plant bioavailable form), oligomeric or polysilicic acid (Iler 1979). Some dissolved silicic acid in the soil solution forms complexes with organic and inorganic compounds. The numerous chains of H_4SiO_4 up to ten silicon atoms in length are classified as the oligomeric or low-molecular-weight-silica, whereas the polysilicic acids with a higher degree of polymerization are the polymeric or the high-molecular-weight-silica (Williams and Crerar 1985). The oligomeric and polysilicic acids are found in chain, branch and sphere forms (Iler 1979). The monosilicic acid form is relevant to plant absorption and nutrition, whereas the polysilicic acid influences soil aggregation. According to Norton (1984), the polysilicic acid links soil particles through the creation of silica bridges that eventually improve soil aggregation, water-holding capacity and buffering capacity, particularly in light-textured soils. Matichencov and Bocharnikova (2001) reported an increase in the water-holding capacity of soils with varying textures (light to heavy) after a month of incubation with silicon-rich materials.

The uncharged H_4SiO_4 is in common soils with pH values <8 (Iler 1979). In most soils and natural waters, the silicic acid is commonly in an undissociated monomeric form (McKeague and Cline 1963; Dietzel 2000). However, the H_4SiO_4 dissociates into $H^+ + H_3SiO_4^-$ at pH values above 9 and further dissociates into $2H^+ + H_2SiO_4^{2-}$ at pH values above 11. For alkaline soils, such as the Solonetz and the Solonchaks, both the undissociated and the dissociated monosilicic acids occur. The formation of stabilized, numerous chains of H_4SiO_4 occurs when the concentration of the silicic acid is high and the pH >9 (Knight and Kinrade 2001). However, significant concentrations of polymerized silicic acid were observed in two acidic, forest soils in Europe, with concentrations as high as 20 % of the total silicon measured in the soil (Wonisch et al. 2008). The stability of the oligomeric form of silicic acid from the dissolution of minerals is short-term and lasts only for a few hours or days under most natural conditions before the breakdown into H_4SiO_4 (Dietzel 2000). The oligomer, polymer and silicon-organic forms of silicic acid are found at high pH values, with the amounts becoming significant at pH values of 11–12 (Iler 1979). The concentration of silicon in soil solutions ranges from 0.09 to 23.4 mg L^{-1}, but the concentrations can be as high as 46.7–93.4 mg L^{-1} in soils with pH values of 10–11 and that contain sodium carbonate (Volkova 1980; Kovda 1985).

The primary sources of H_4SiO_4 in the soil solution are the various forms of silicon dioxide, silicate minerals and plant residuum. The amount of H_4SiO_4 released by the various forms of SiO_2 is dependent on the physico-chemical properties. The SiO_2 in the soil influences the concentration of the H_4SiO_4 in the soil solution. Those forms that occur as nepheline, diopside, and augitein in a dispersed state may supply between 7 and 9 mg silicon L^{-1}, whereas the bioptide, microline and labradoride may supply between 2.3 and 3.5 mg silicon L^{-1}. However, quartz has a low solubility rate and releases only 1.6–1.9 mg silicon L^{-1} (Keller 1955; Lindsay 1979; Drees et al. 1989). The weathering of the silicate minerals releases silicon into the soil solution, which can be combined with other elements to form clay minerals, be released into the streams and the oceans or be used for uptake by plants and microorganisms. A small amount of silicon is contributed to the soil solution by minerals that are insoluble and resistant to weathering, which include feldspar and a number of complex silicates such as circone, garnet and tourmaline (Kovda 1985).

The amount of H_4SiO_4 in the soil solution is affected by many factors and the solubility of silicon containing minerals is affected by pH, temperature, particle size, water and organic matter contents, and redox potential (Savant et al. 1997). Overall, the soil pH regulates the solubility and the mobility of silicon. The adsorption-desorption processes affect the concentration of H_4SiO_4 in the soil solution and are very dependent on the soil pH (McKeague and Cline 1963). The maximum adsorption of H_4SiO_4 occurs at a pH of 9–10, and at pH values below or above these levels, the amount of adsorption is reduced. The adsorption, polymerization and coagulation of H_4SiO_4 in saline soils are high (Brown and Mahler 1988). The amount of adsorbed H_4SiO_4 also increases in soils that contain large amount of allophanes, Fe-enriched crystal minerals, and particularly, the more reactive hydroxides of multivalent metals. The production of silicon dioxide (SiO_2) deposits in the form of crusts is enhanced during the evaporation, transpiration and freezing processes (McKeague and Cline 1963). The application of acid-producing fertilizer increases the concentration of H_4SiO_4 in the soil solution, whereas liming and high organic matter content result in a reduction in the concentration and mobility of the H_4SiO_4 (Panov et al. 1982; Allmaras et al. 1991). The alkalinized H_4SiO_4 can be redeposited as a cementing and a blocking agent in the lower horizons of the soil profile.

The concentration of H_4SiO_4 in the soil solution also changes seasonally within ecosystems. In grassland ecosystems, the maximum concentration of H_4SiO_4 is observed during the spring and summer when the temperature favors biological activity (Volkova 1980; Bystritskaya 1987; Fernandes and Macias 1987). However, in forests, the highest concentration of H_4SiO_4 was observed during the autumn leaf fall (Volkova 1980; Pervova and Evdokimova 1984).

3. Silicon Adsorbed on Solid Phases The fractions of dissolved silicic acid in the soil solution are adsorbed onto a variety of solid phases in soils, including clay particles and Fe and Al hydroxides (Hansen et al. 1994; Dietzel 2002). A minimal reduction in the concentration of silicon in the soil solution is attributed to the adsorption by secondary clay minerals (Siever and Woodford 1973). However, the Fe and Al hydroxides have strong adsorption capacity, which can remove significant amounts of dissolved silicon from the soil solution (Beckwith and Reeve 1963; McKeague and Cline 1963; Cornell and Schwertmann 1996).

The pH, soil redox potential (Eh), and the type of metal influence the adsorption of monosilicic acid by oxides. The amount of monosilicic acids that is adsorbed by oxides increases from pH 4 to pH 9, and the amount is notably higher when the metal oxides in the soil are Al-based rather than Fe-based. Ponnamperuma (1965) reported that with the increased submergence time of soil the corresponding reduction in the Eh was accompanied with an increase in the solubility of the soil silicon. This increase in silicon in the soil solution was attributed to the release from ferri-silica complexes under anaerobic soil conditions. The Al hydroxides are more effective than the Fe oxides in adsorbing the H_4SiO_4 in the soil solution (Jone and Handreck 1963, 1965, 1967; McKeague and Cline 1963). In general, the silicic acid is adsorbed onto secondary Fe-based oxides; a higher amount of silicic acid is adsorbed on the short-range, ordered ferrihydrite than that on the crystalline goethite (Delstanche et al. 2009). The OH group of the Fe-oxide surface is replaced with the H_4SiO_4 through ligand exchange, which eventually forms a silicate bi-dendate innersphere complex (Parfitt 1978; Pokrovsky et al. 2003; Hiemstra et al. 2007). The polysilicic acid is also formed through specific interaction of the Fe-oxide surface with the orthosilicic acid (Dietzel 2002). The iron oxides are commonly found in soils, and therefore even if the silicon adsorbing capacity is less effective compared with the Al oxides, the iron oxides will control, to some degree, the concentration of H_4SiO_4 in the liquid phase (McKeague and Cline 1963; Schwertmann and Taylor 1989; Opfergelt et al. 2009).

Silicon Cycle in Soil

The solid, liquid, and adsorbed phases of silicon are the key components of the silicon cycle in soil (Fig. 2.2). The liquid silicon phase consists of H_4SiO_4 and the polymerized and complexed silicic acid in soil solution, and the uncharged form of H_4SiO_4 is the only form that is absorbed by plants and microorganisms. The absorbed silicon is later deposited as polymerized silica within the plant tissues or the cell structure of the microorganisms. These polymerized silica bodies return to the topsoil in the litter fall and the remains of microorganisms and eventually enter the highly soluble biogenic silica pool that contributes to the silicon in the soil solution (Drees et al. 1989; Van Cappellen 2003; Farmer et al. 2005; Saccone et al. 2007; Fraysse et al. 2010). Conley (2002) estimated that 60–200 Tmol silicon per year is stored in plants. Silicon is also added to soils with applications of manure and compost, and the decomposition of silicon-rich manure can increase the level of available soil silicon (Song et al. 2013). The silicon rarely interacts with dissolved organic matter but does form colloidal aluminum-silicon polymers (suspended silicon particles) at many soil solution pH values (Doucet et al. 2001). The chemistry

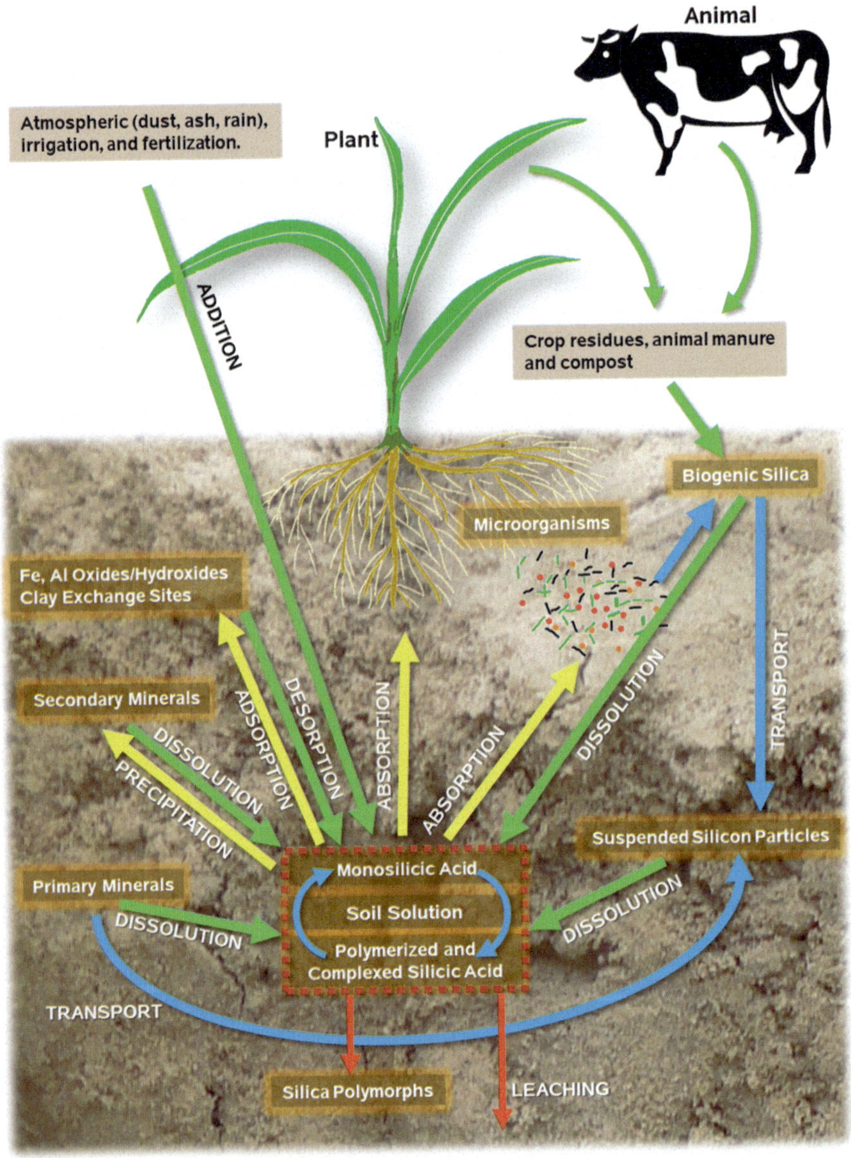

Fig. 2.2 Comprehensive cycle of silicon in soil (*Green arrows* represent transformation or processes which raise silicon concentration in soil solution. *Yellow arrows* represent the transformation or processes which reduce silicon concentration in soil solution. *Red arrows* represent processes that result in silicon loss from the soil system or production of stable, plant unavailable form of silicon. *Blue arrows* represent transformation processes of silicon into a silica pool that contributes this element into the soil solution)

of silicon in the liquid phase is regulated by a number of processes: (a) the dissolution of silicon that contains primary and secondary minerals, (b) the absorption of H_4SiO_4 in the soil solution by the vegetation and microorganisms, (c) the silicon adsorption on and the desorption from various solid phases, (d) the preservation of the stable silicon in the soil profile (silica polymorphs), (e) leaching, and (f) addition (i.e., fertilization, irrigation, atmospheric, plant litter, animal manure, and remains of microorganisms). The natural waters used as irrigation may contain different forms of silicon, including ionic, molecular, and aggregate silicon. Silicon is also added to the soil in atmospheric deposition via wind-blown dust and phytolith particles from savanna fires (Kurtz et al. 1987; Street-Perrott and Barker 2008; Opfergelt et al. 2010). However, the contribution of silicon to the soil solution from the atmosphere is very low compared with the other silicon inputs to the soil-plant system (Street-Perrott and Barker 2008).

Interaction of Silicon with Other Plant Nutrients in the Soil

The application of a silicon-rich material influences the dynamics of different elements in the soil. The outcome of the reactions maybe beneficial (e.g., alleviate heavy metal toxicity) but may also be negative with the reduced availability of several plant-essential nutrients. The ability of silicon to influence the dynamics of elements in the soil is because of the high adsorptive capacity of the form of the silicon-rich materials that are commonly found in and added to the soil. Additionally, an increased concentration of silicic acid ions in the soil solution leads to the formation of complexes with heavy metals in the soil and to competition with other ions for adsorption sites.

The nutrients in the soil solution with a positive charge are adsorbed onto a silica surface. In a study conducted by Tokunaga (1991), the leaching losses of K and other mobile nutrients from the surface soil horizon were reduced because of a silica surface. The nutrients adsorbed onto the silica surface remain available to plants and formed the basis for slow-release fertilizer technology (Volker et al. 1985; Komisarov and Panfilova 1987).

According to a plethora of published information, phosphate availability increases following silicon fertilization (Gladkova 1982; Singh and Sarkar 1992; O'Reilly and Sims 1995; Matichenkov and Ammosova 1996). Matichenkov and Ammosova (1996) and Lindsay (1979) outlined the series of reactions involved between the silicate and the phosphate ions in which in the final reaction resulted in the release of phosphates into the soil solution (Eqs. 2.1a, 2.1b, and 2.1c). The fertilization with silicon increased the amount of dissolved silicon in the soil solution (H_4SiO_4), and the amount of silicon adsorbed onto the slightly soluble phosphates of Al, Ca, ferric and Mg was followed by the desorption of the phosphate anion.

$$CaHPO_4 + Si(OH)_4 = CaSiO_3 + H_2O + H_3PO_4 \tag{2.1a}$$

$$2Al(H_2PO_4)_3 + 2Si(OH)_4 + 5H^+ = Al_2Si_2O_5 + 5H_3PO_4 + 5H_2O \quad (2.1b)$$

$$2FePO_4 + Si(OH)_4 + 2H^+ = Fe_2SiO_4 + 2H_3PO_4 \quad (2.1c)$$

Notably, a silicon fertilizer has the capacity to adsorb the dissolved phosphates in the soil solution, including those released from the exchange reaction between the silicate and the phosphate ions. In one experiment, Matichencov and Bocharnikova (2001) showed that a silicon source in the form of steel slag was the most effective in the adsorbing of phosphate in solution compared with amorphous fine SiO_2, calcium carbonate, calcium silicate, and the industrial silicon by-product from the electric production of P. The steel slag consistently adsorbed >95 % of the phosphate in solution, whether the phosphate concentration in solution was as low as 0.5 mg P L^{-1}or as high as 10 mg P L^{-1}. However, the amount of phosphate adsorbed by the other silicon sources increased significantly with the increase in the phosphate concentration in solution. For example, the amorphous SiO_2 adsorbed 2 %, 4 %, and 52 % of the 0.5, 2, and 10 mg P in a liter of solution, respectively.

In the soil system, the relationship between the phosphate and the H_4SiO_4 is antagonistic; the amount of phosphate ion that is released into the soil solution increases with increasing concentrations of H_4SiO_4. The antagonistic reaction between the phosphate and the H_4SiO_4 ions is explained by the strong competition for specific sorption sites (Brown and Mahler 1987). However, Jones and Handreck (1967) noted that this competition is more likely a long-term effect of the silicic acid; for example, gibbsite, when silicified into kaolinite, has reduced affinity for phosphate ions. The short-term competition between the silicic acid and the phosphate ions for adsorption sites has a minimal contribution, or possibly none, to the concentrations in the soil solution. The silicic acid is attracted to the hydrogen bond of an oxygen atom that bridges two metal atoms, whereas the phosphate (basic) is attracted to the metal atoms; these two sites are different types. The high P sorption capacity of a low pH soil from the coastal plain of Georgia was markedly reduced with the application of sodium silicate, an effect attributed to the increase in soil pH by Owino-Gerroh and Gascho (2004). These authors noted that the amorphous silicic acid (from silicate ions) had a lower negative surface charge than that of the phosphate ion. Thus, when these two ions are present in the soil solution, the amorphous silicic acid is preferentially adsorbed over the phosphate ion. Earlier propositions were also considered and include the following: (1) the increase in alkalinity caused by the increase in the concentration of monosilicic acid liberated the phosphate with the dissolution of iron and aluminum oxides, and (2) the monosilicic acid lowered the activity of aluminum ions in solution by preventing these ions from precipitating the phosphate (Jones and Handreck 1967).

The application of the silicon rich materials fly ash and steel slag resulted in an increase in the soil pH (>1 unit), which decreased the phytoavailability of Cd, Cu, Pb and Zn by 60 % and eventually reduced the uptake of the heavy metals in rice (Chen et al. 2000; Gu et al. 2011). Additionally, the heavy metal diffusion fluxes from the soil to the solution were reduced by 84 % because of precipitations with silicates, phosphates, and hydroxides. Moreover, the solubility of these heavy metal silicates was very low (Schindler et al. 1976). A recent study conducted by Tubaña et al. (2012a) showed that the addition of increasing rates of steel slag resulted in a steady decline in the concentrations of the Mehlich-3 extractable Fe and Ni, and the decline was attributed to the increase in soil pH as the application rates of the applied steel slag were also increased. Wallace (1993) explained that despite the high soluble Fe content in the anaerobic soils in which paddy rice is grown, the high concentration of silicon in rice creates an alkaline rhizosphere that decreases the availability of Fe. However, the heavy metal content in the soil solution also increased when the concentration of the H_4SiO_4 was increased (Schindler et al. 1976; Bocharnikova et al. 1995). This result was caused by the ability of the H_4SiO_4 ion (even at low concentrations) to form slightly soluble complexes with heavy metals. However, at a high concentration of H_4SiO_4 in the soil solution, the heavy metals are immobilized by the precipitation of silicates, which leaves a low concentration of soluble silicates for plant uptake (Jones and Handreck 1967; Lindsay 1979; Snyder et al. 2007). Ma and Yamaji (2006) noted that the silicon in soil becomes unavailable for plant uptake when it forms silicates or oxides with other compounds.

Earlier studies showed that the application of silicon-rich materials effectively reduced the Al toxicity in plants through the reduced uptake of Al (Haak and Siman 1992; Myhr and Estad 1996). The potential mechanisms for this effect include the following: (1) the precipitation of Al caused by the increased soil pH as a result of the elevated concentration of H_4SiO_4 (Lindsay 1979); (2) the H_4SiO_4 was adsorbed on Al hydroxides, which formed a less mobile compound and diminished the activity of the phytotoxic Al in solution (Panov et al. 1982; Baylis et al. 1994); and (3) the mobile Al was strongly adsorbed on the silica surfaces (Schulthess and Tokunaga 1996). The reduction in Al toxicity to plants was not caused entirely by the immobilization of Al in the soil or growth media. Rahman et al. (1998) reported that an increase in silicon nutrition increases the tolerance of the plant to excessive amounts of absorbed Al. Similarly, Liang et al. (2005a) showed that the silicon-enhanced tolerance of corn to Cd toxicity was attributed to both the Cd immobilization caused by the increase in soil pH and the silicon-mediated detoxification of the Cd in the plant. In rice, the oxidation of ferrous to ferric ion is increased because of an increase in the silicon-induced oxidizing capacity of the roots (Ma and Takahashi 2002). The ferrous form of iron is preferred for plant uptake compared with the ferric form, which prevents the excessive accumulation of Fe in flooded rice. Wallace (1993) suggested that silicon increased the release of OH^- from the roots, and the increase in the soil pH eventually led to the decrease in the solubility of Fe. Unlike the *in planta* mechanisms (the internal silicon-mediated mechanisms in plants; Table 2.1), the silicon-mediated mechanisms involved in the prevention of excessive uptake of metals from the soil and the roots require further study (Kirkham 2006).

Table 2.1 Internal silicon-mediated mechanisms involve in enhancing the plant's tolerance to heavy metal toxicity

Heavy metal	Crop	References	Mechanisms
Aluminum	Barley	Hammond et al. 1995	Exclusion of Al from the subtending tissue as a result of silicon deposition at the epidermis, restricting total overall Al uptake into the root
	Corn	Wang et al. 2004	Formation of hydroxyaluminosilicates in the apoplast of the root apex reducing the mobility of apoplastic Al
		Kidd et al. 2001	Mediates the metabolism of flavonoid-phenolic compounds which strongly chelate Al
Arsenic	Rice	Seyfferth and Fendorf 2012	Silicon competes with arsenate ions for root entry points
Cadmium	Corn	Liang et al. 2005a	Co-precipitation of Cd with silicates resulting in strong binding of Cd to cell walls thereby reducing the concentration of Cd in cytosols or symplast
		Wang et al. 2000	Formation of colloidal silicon in cell walls which has high specific adsorption property to Cd preventing Cd uptake into the cell
		Cunha and Nascimento 2009	Structural alterations on xylem diameter, mesophyll and epidermal thickness, and transversal area occupied by collenchyma and midvein; deposition of silica in the endodermis and pericycle of roots
	Rice	Nwugol and Huerta 2008	Cell wall-bound silicon inhibit apoplastic Cd uptake by covalently bonding with Cd and trapping Cd as it diffuses through the cell wall and intracellular spaces.
	Peanut	Shi et al. 2010	Increased activities of antioxidant enzymes; inhibition of Cd transport from roots to shoots possibly due silicon-mediated changes on cell wall properties and competition for uptake sites
Lead	Cotton	Bharwana et al. 2013	Enhanced the activities of major antioxidant enzymes preventing plant tissue from membrane oxidative damage
	Cowpea	Iwasaki et al. 2002a	Enhanced adsorption of Mn on cell walls reducing the amount of soluble apoplastic Mn
		Iwasaki et al. 2002b	Interaction of silicon with phenolic substances maintains the apoplast in reduced state preventing the oxidation of Mn by guaiacol-peroxidase
	Cucumber	Rogalla and Römheld 2002	Strong binding of Mn to cell walls and a lowering of the Mn concentration within the symplast
		Shi et al. 2005	Enhanced production of enzymatic (e.g. superoxde dismutase, ascorbic peroxidase, glutathione reductase) and non-enzymatic (e.g. ascorbate and glutathione) antioxidants resulting in reduced membrane lipid peroxidation

(continued)

Table 2.1 (continued)

Heavy metal	Crop	References	Mechanisms
Zinc	Corn	Kaya et al., 2009; Neumann and zur Nieden 2001	Formation of less soluble zinc-silicates in cytoplasm
		Cunha and Nascimento 2009	Structural alterations on xylem diameter, mesophyll and epidermal thickness, and transversal area occupied by collenchyma and midvein; deposition of silicon in the endodermis and pericycle of roots

Silicon in Plants

1. Silicon Uptake, Transport and Deposition in Plant Plants uptake silicon from the soil solution in the form of H_4SiO_4, which is commonly found at concentrations that range from 0.1 to 0.6 mM at the pH levels found in most agricultural soils (Knight and Kinrade 2001). According to Ma et al. (2001a), the lateral roots of rice are involved in the uptake of silicon. Cornelis et al. (2011) described the different mechanisms by which the silicon is absorbed by plants, i.e., active, passive and rejective. The amount of uptake of silicon by the active mechanism is typically larger than that predicted based on the mass flow and is attributed to the density of silicon transporters in the roots and shoots that facilitate the absorption process across the membranes of root cells. In rice, the transporters mediate both the radial transport and the xylem loading of silicon (Mitani and Ma 2005). Moreover, these transporters were recently identified and were coded by low-silicon genes such as the *Lsi1* and *Lsi2* in roots and the *Lsi6* in shoots (Mitani and Ma 2005; Ma et al. 2006, 2007; Yamaji et al. 2008). The *Lsi1*may encode a membrane protein similar to the water channel proteins, also known as aquaporins (Ma et al. 2006). The amount of uptake of silicon by the plant via the passive mechanism is likely entirely driven by mass flow. In the rejective mechanism, the buildup of the concentration of H_4SiO_4 in the soil solution typically results from the low concentrations of silicon that are absorbed by plants.

Takahasi et al. 1990 categorized plant species based on the mechanisms of silicon uptake. The plants that rely primarily on active, passive or rejective mechanisms are classified as high-, intermediate- or non-accumulators, respectively. The plants in the high-accumulator category have a silicon content in the shoot that ranges from 1.0 % to 10 % dry weight and are primarily monocotyledons such as bamboo (*Bambuseae*), barley (*Hordeum vulgare*), rice (*Oryza sativa*), sorghum (*Sorghum bicolor*), sugarcane (*Saccharum officinarum*), and wheat (*Triticum aestivum*) (Liang et al. 2007; Ma et al. 2001b; Ma and Takahashi 2002). Because of the efficient silicon uptake system of the high-accumulators, the amount of silicon uptake by the plant from the soil is several times higher than the uptake of some of the essential macro- or micronutrients. For example, the uptake of N is the largest among the essential nutrients, but the accumulation of silicon may be twice the amount of N in rice.

The intermediate-accumulator plants are mostly dryland Gramineae with shoot silicon contents that range between 0.5 % and 1.5 % dry weight. The dicots, which accumulate <0.2 % shoot dry weight silicon, form the low-accumulator group. Mitani and Ma (2005) attributed the low silicon accumulation in this group of plants to a lack of specific transporters to facilitate the radial transport and the xylem loading of silicon and suggested that the transport of silicon across cells was accomplished via a passive diffusion mechanism. Later, Liang et al. (2006) showed that both the active and the passive uptake of silicon, which occur in high-accumulator plants, are also found in the intermediate-accumulator plants (e.g., sunflower and wax gourd).

The absorbed H_4SiO_4 is transported through the xylem and is deposited in the leaf epidermal surfaces in which it is condensed into a hard, polymerized silica gel ($SiO_2{\cdot}nH_2O$), also known as a phytolith (Yoshida et al. 1962; Jones and Handreck 1965, 1967; Raven 1983). The absorbed H_4SiO_4 is preferentially deposited in the abaxial epidermis, but as the leaf grows, the deposition occurs in the epidermis (Hodson and Sangster 1988). In wheat, the silicon is in all tissues but high concentrations are found in the inner tangential and radial walls of the endodermis (Bouzoubaa 1991). The phytoliths are found in specific cells, the silica cells, which are in vascular bundles and in silica bodies in bulliform cells, fusoid cells or prickle hairs in rice, wheat, and bamboo, respectively (Dietrich et al. 2003; Motomura et al. 2004; Ma and Yamaji 2006). According to Lanning (1963), the phytoliths are best classified as biogenic opal (Si-O-Si bonding). The SiO_2 precipitation in plants occurs at concentrations of H_4SiO_4 greater than 2 mol m^{-3} (Osuna-Canizales et al. 1991) and occurs primarily in the epidermis of the shoots, in addition to the vascular system and the endodermis of roots of some plant species (Raven 1983; Lux et al., 2003a, b). The deposited silica is immobile and is not transferred to actively growing or meristematic tissues (Elawad and Green 1979; Ma et al. 1989; Epstein 1999). Transpiration remains a viable option as one of the primary drivers in silicon transport and deposition in plants, and therefore, the duration of plant growth significantly affects the concentration of silicon; for example, older leaves contain more silicon than younger leaves (De Saussure 1804; Henriet et al. 2006). Based on earlier research, the $SiO_2.nH_2O$ framework possibly binds with organic components (Lanning 1963). Conversely, the studies by Casey et al. (2003) and Ma et al. (2004) confirmed that only the mono- and the di-silicic acids but not the organosilicic complexes were found in the xylem exudates of rice and wheat.

2. Effects on Plant Growth Silicon is a known essential nutrient to only two groups of plants, i.e., the scouring rushes and the diatoms and other members of the yellow-brown or golden algae (Epstein 1999). To date, whether silicon is essential for higher plants remains uncertain because no evidence to demonstrate the direct involvement of silicon in plant metabolism has been found and no silicon-bearing organic compound has been identified in higher plants (Ma et al. 2001b; Knight and Kinrade 2001; Ma and Takahashi 2002; Richmond and Sussman 2003). However, the amount of literature that documents the benefits of silicon on the growth of a wide variety of agronomic and horticultural crops is vast and continues to increase. The beneficial effects of silicon become more evident when plants are in stressed (biotic or abiotic stress) environments than in those growing under optimal conditions (Li et al. 2007; Epstein 1999; Bélanger et al. 1995; Datnoff et al. 1997). The beneficial effects of silicon on plant growth and development are based on several mechanisms, which include the formation of a protective outer layer composed of silica deposits, the reactivity of the absorbed silicon with the heavy metals ions and other compounds within plants and the metabolic functions of silicon in stressed plants.

2.1. Reinforced Plant Protective Layer and Mechanical Structure In the amelioration of biotic-related stresses, the role of silicon was first recognized in the modification of plant cell wall properties (Horst et al. 1999; Fawe et al. 2001; Lux et al. 2002; Iwasaki et al. 2002a, b). The deposition of biogenic silica in shoots increases the structural component of the plant and creates a hard outer layer (Rafi et al. 1997; Bélanger et al. 2003). Most of the reported benefits in crop quality and yield following silicon fertilization resulted from the improved overall mechanical strength and an outer layer of enhanced protection for the plant (Epstein 1999, 2001; Ma and Takahashi 2002; Epstein and Bloom 2005).

The silicon-enhanced mechanical defense of plants significantly reduces the damage caused by insects and grazing animals. For example, wild rabbits (Cotterill et al. 2007) and locusts (Hunt et al. 2008) preferred to eat unfertilized grasses compared with silicon-fertilized grasses. Savant et al. (1997) reported that silicon fertilization in rice reduced the damage caused by insect borers, yellow borers, rice chlorops, rice leafhoppers, brown leafhoppers, and mites. Gomes et al. (2005) attributed the reduction in aphid damage to the decreased number of aphids that were observed in infested plants fertilized with silicon. According to Goussain et al. (2005), the silicification did not create a physical barrier against penetration of the stylus of aphids, but they did observe a chemical-induced removal of the stylus, which eventually reduced the amount of sap consumed by the aphids.

Reports also indicate that silicon fertilization improved the tolerance of plants to stress from the lack of moisture (Janislampi 2012; Rizwan et al. 2012). The silicon fertilized crops maintained higher biomass and grain yields with a deficiency of water (Eneji et al. 2005, 2008; Pei et al. 2010). The wheat plants treated with silicon fertilizer under drought stress had higher stomatal conductances, relative water contents, and water potentials than nontreated plants (Pei et al. 2010). The reduction in water loss through transpiration (Hattori et al. 2005) and the decreased uptake of water (Eneji et al. 2005) were attributed to the larger and thicker leaves of silicon-treated plants and to the higher silicon deposition in the cell walls of epidermal tissues (prevents excessive water loss through transpiration) and the xylem vessels (prevents compression of the vessels) than nontreated plants. The thickened silicate layer on the leaf surface also reduces cuticular transpiration. Thus, the silicon increased the drought tolerance of plants not only by maintaining water balance, photosynthetic efficiency, erectness of plant canopy structure, and structure of the xylem vessels under high transpiration rates (Hattori et al. 2005), but also improved the development of secondary and tertiary cells of the endodermis for a better root resistance to dry soils and a faster growth of roots to explore a larger volume of soils than plants not treated with silicon (Hattori et al. 2003, 2005). Ma et al. (2001b) also reported an increase in the resistance of rice to typhoon damage, which was attributed to the increase in rigidity with the silicification of shoots.

2.2. Reactivity of Silicon with Other Elements and Compounds Inside the Plants According to Cocker et al. (1998), the beneficial effects of silicon in plants are based on two aspects, i.e., solution chemistry and *in planta* mechanisms. These authors described the co-deposition of silicon and Al that formed less soluble aluminosilicates or hydroxyaluminosilicates within the root cell wall as responsible for the reduced concentration of free, toxic Al^{3+} ions in plants. A more recent and comprehensive review of the silicon-mediated mechanisms used to alleviate the abiotic stress caused by heavy metal toxicity, salinity, drought and freezing was conducted by Liang and his colleagues (2007). These authors grouped the mechanisms for the alleviation of metal toxicity with the increased level of silicon within the plant into two groups, external and internal. The external mechanisms are characterized by the inhibition of the absorption of metal ions by plants through the following processes: (1) the reduction in metal activity via increased ionic strength or pH, (2) the formation of metal-phenolic complexes caused by the silicon-mediated release of phenolic compounds, and (3) the co-deposition between the silicon and the metal ions in growth media. In contrast to the external mechanisms, the internal mechanisms occur within the plant and involve the following processes: (1) the enhancement of the antioxidant systems in the plant, (2) the complexation or co-precipitation of metal ions with silicon, (3) the uptake processes, and (4) the compartmentalization of metal ions. The changes in the plant cell wall properties not only contributed to the mechanical strength in the *Gramineae* but also inhibited the transport of metals (Cunha and Nascimento 2009). According to the authors, the reduction in metal transport from the roots to the shoots may have resulted from the thickening of the Casparian strips in the endodermis and the cell wall of the xylem and the pericycle, in addition to the deposition of lignin (endodermis, epidermis and exodermis) and silicon (endodermis) in the cell walls (Shi et al. 2005; Cunha and Nascimento 2009).

Many studies demonstrated that silicon fertilization of several types of crops reduced the metal uptake and toxicity. However, the mechanisms for the alleviating action of silicon on metal toxicity were not determined in all of the studies, including those on the toxic effects of Cu on spring wheat (Nowakowski and Nowakowska 1997) and of As on rice (Guo et al. 2005). Several silicon-mediated mechanisms to alleviate heavy metal toxicity in a wide array of crops were reported in the literature (Table 2.1). Silicon is generally reactive to heavy metals and impairs the translocation inside the plants and eventually reduces the toxic effect to the plant (Rahman et al. 1998; Neumann and Nieden 2001; Richmond and Sussman 2003; Ma et al. 2004). The reduced translocation of absorbed heavy metals in plants was attributed to the buildup of silica deposits in the cell walls that bound the metal ions and prevented the distribution of the ions from the roots to the shoots, in addition to the complex formation of silicon with metal ions that limited the translocation to different parts of the plants (Gu et al. 2011; Ma et al. 2001b). In cucumber, the binding of Mn to the cell walls resulted in decreased Mn content in the symplasts (Rogalla and

Romheld 2002), whereas a similar mechanism was reported for Cd in peanut and Al in barley (Baylis et al. 1994; Shi et al. 2010). The formation of an aluminum-silicon complex eventually prevented the penetration of Al into the root cortex of sorghum (Liu et al. 2004). The plants that suffer from heavy metal toxicity may benefit from silicon application through the increased release of compounds that immobilize the heavy metal ions. Additionally, the release of a phenolic compound was associated with the silicon-mediated increased resistance to Al in an Al-resistant maize cultivar (Kidd et al. 2001).

Silicon also reduced the oxidative stress induced by B (semi-heavy metal) toxicity (Gunes et al. 2007; Inal et al. 2009). Additionally, silicon inhibited the accumulation of Na in salt-stressed plants through a silicon-induced reduction of the transpiration rate and a partial blockage of the transpirational bypass flow (Matoh et al. 1986; Yeo et al. 1999) and a silicon-induced stimulation of the root plasma membrane H^+-ATPase (Liang 1999; Liang and Ding 2002; Liang et al. 2005, 2006). In the latter study, an eventual reduction in the Na content in the shoots of barley was the result of an increase in the uptake and transport of K and a decrease in the uptake and transport of Na from roots to shoots. In addition to altering the structure, integrity, and functions of the plasma membrane, silicon alleviates the problems associated with salinity with a reduction in the stress-dependent peroxidation of membrane lipids through the stimulation of antioxidant enzyme and nonenzyme activities in the plants (Liang et al. 1996, 2003, 2005, 2006; Liang 1999). These observations were consistent with the research conducted on several intermediate- or low-silicon accumulator plants, such as cucumber (Zhu et al. 2004) and tomato (Al-Aghabary et al. 2004).

As described above, silicon fertilization also alleviates problems associated with moisture stress in plants. Gong et al. (2005) and Pei et al. (2010) documented the benefits of silicon fertilization to drought-stressed plants at the metabolic level in wheat. In Gong et al. (2005), in silicon-treated wheat plants under moisture stress, a corresponding increase in the antioxidant defenses helped to maintain physiological processes such as photosynthesis. The improvement in the growth of wheat under short-term water stress when supplied with a silicon fertilizer was attributed to an enhancement of the antioxidant defense system rather than to the adjustment in the osmotic pressure.

Silicon Sources

From a global perspective, Guntzer et al. (2012) highlighted the important role of silicon in the maintenance of crop productivity. According to Guntzer et al. (2012), among the top ten most produced crops worldwide, seven of these crops are silicon accumulators, which include maize, rice, sugar beet, sugarcane, and wheat. The estimated amount of silicon removed annually by the different agricultural crops on a global scale is between 210 and 224 million tons (Bazilevich 1993; Reimers 1990; Savant et al. 1997). For the high silicon-accumulator crops (e.g., rice, sugarcane, and wheat), the removal of silicon from the soil is significantly higher than the removal in natural systems. For example, for sugarcane and rice, the silicon removal rates were between 300 and 500 kg ha^{-1} $year^{-1}$, respectively, compared with the US grasslands that averaged only between 22 and 67 kg ha^{-1} $year^{-1}$ (Meyer and Keeping 2001; Blecker et al. 2006; Makabe et al. 2009). With years of continuous and intensive cropping, the harvest of silicon-accumulator crops results in a significant reduction in the amount of plant-available silicon in soils (Meunier 2003; Meunier et al. 2008). Desplanques et al. (2006) noted that if the rice production in the fields of Camarque relied entirely on amorphous silica as the source of silicon, the reserve of plant-available silicon would be exhausted after five years of cultivation. According to Hodson et al. (2005), the concentration of silicon in plants depends primarily on the phylogenetic position of the plant, compared with the environmental effects that encompass silicon concentrations in the soil and soil solution and the pH. Nevertheless, various amounts of silicon uptake are reported for a given plant species; thus, although silicon accumulation is primarily a phylogenetic feature, the amount of plant-available silicon in the soil affects the amount of silicon that is absorbed by the plant (Deren et al. 1992; Ma and Takahashi 2002; Henriet et al. 2006).

The processes that regulate the concentration of silicon in the soil solution occur immediately to replenish the silicon that is removed by plants until equilibrium is reached between the liquid and the solid phases of silicon (Fig. 2.2). The soils with high buffering capacity (e.g., recent volcanic soils) easily replenish the lost silicon and maintain high levels of dissolved silicon for plant uptake. However, the removal of silicon from some types of soil (e.g., highly weathered, organic, and intensively cropped) may require some time to replenish, even with accelerated mineral weathering, depolymerization of polysilicic acid, and dissolutions of silicate complexes with heavy metals, hydroxides and organic matter; thus, these types of soils require the addition of silicon through fertilization with Si-rich materials.

The purpose of silicon fertilization is to increase the concentration of H_4SiO_4 in the soil solution. Matichencov and Bocharnikova (2001) provided an overview of the formation of the different silicic acid species in soil solution as affected by the rates of silicon fertilization. Three phases were established based on the changes in the concentrations of monosilicic and polysilicic acids (Fig. 2.3). Phase A occurs at the low end of the range of silicon fertilization rates for which the concentration of the H_4SiO_4 in the soil solution increases. As the rate of added silicon increases, the concentration of the monosilicic acid reaches a certain point and then begins to polymerize (the formation of polysilicic acid). The concentration of the silicon in the soil solution ranges between 0.01 and 1.99 mM silicon (Karathanasis 2002). Tan (1994) and Matichenkov and Ammosova (1994) further reported that polymerization occurs when the silicon concentration in the soil solution exceeds 65 mg L^{-1}. At this concentration and above, a mixture of H_4SiO_4, polysilicic acid and silicon-organic compounds is found in the soil solution, which indicates that ~2 mM silicon is potentially the concentration at which polymerization begins (phase B). During phase B, the H_4SiO_4 from the addition of silicon fertilizer produces polysilicic acid. Thus, even when the amount of silicon added to the soil is increased, the level of the

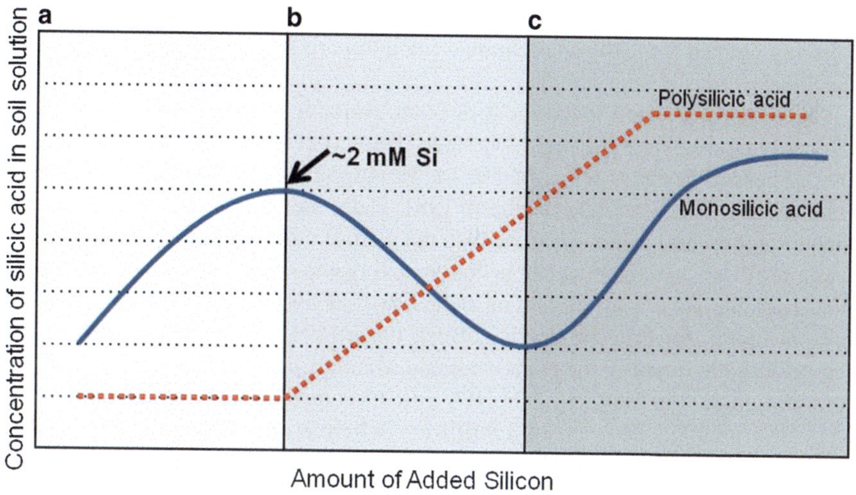

Fig. 2.3 Monosilicic and polysilicic acid fractions in the soil solution as affected by the amount of added silicon. Letters on the x axis represent a range of silicon fertilization rates where A is the low end, B is the middle and C is the high end rate (Adapted from Matichencov and Bocharnikova (2001))

H_4SiO_4 continuously declines, and the amount of polysilicic acids increases. In phase B, the effect of silicon fertilization is evident only in the amount of polysilicic acid. Phase C is characterized by both the synthesis of the polysilicic acids and the increase in the H_4SiO_4 concentration in the soil solution, within the range of rates of added silicon to the soil. Within phase C, both the polysilicic and H_4SiO_4 acids increase in concentration with the increased rate of added silicon. Notably, the processes that occur within these three phases (Matichencov and Bocharnikova 2001) are exclusively dependent on the concentration of silicic acid (because of the silicon addition). Therefore, the influences of pH, temperature, and the concentration of heavy metal ions were not included in this illustration (Fig. 2.3); however, these factors have a strong influence on the stability of both the H_4SiO_4 and the polysilicic acids in soil solution (Yates et al. 1998).

Calcium silicate occurs as prismatic crystals of wollastonite (Maxim et al. 2008), and pulverized wollastonite is commonly used in many silicon studies because of the high content of calcium silicate (at least 50 % SiO_2). The deposits of wollastonite are not typically found in the pure form (calcium silicate), and therefore, labor-intensive, expensive refining processes are required, which limit the mass production of wollastonite as a fertilizer (Park 2001; Maxim et al. 2008). Materials, such as magnesium silicate, contain large amounts of silicon, but are not considered a suitable source for silicon fertilizer because of the poor solubility (Weast et al. 1985). Currently, the silicon-containing industrial by-products or slags are most commonly used as silicon fertilizers. These industrial by-products, such as those from the electric production of P and the production of steel/iron, are inexpensive and accessible sources of silicon for the production of crops. The silicate slags often contain a small fraction of easily soluble silicon (Gascho 2001) but have the added benefit as liming agents, typically with similar calcium carbonate equivalents (Heckman et al. 2003).

The composition and the amount of plant-available silicon found among these silicon-containing slags are highly variable (Datnoff et al. 2001; Ma and Takahashi 2002). These differences are caused by the variation in the speed of cooling and the granular size of the material (Takahashi 1981; Datnoff et al. 1992). Additionally, the silicate slags are more cost-effective than the wollastonite. Thus, for the purposes of silicon fertilizer management and economics, it is important to know the amount of plant-available silicon in the silicon-rich industrial by-products that are commercially available for crop production. Buck et al. (2011) evaluated several of the methods used to quantify the plant-available silicon from the industrial by-products (solid or liquid). The $Na_2CO_3+NH_4NO_3$ extraction method was optimal to estimate the plant-available silicon in solid fertilizers, whereas quantifying the total silicon content via HCl+HF digestion was suitable for liquid fertilizers. Recently, the 5-day Na_2CO_3-NH_4NO_3 soluble silicon extraction method was recognized as the official method in the United States and was approved by the Association of American Plant Food Control Officials to quantify the plant-available silicon in solid fertilizer products (Sebastian 2012; Sebastian et al. 2013). This method originated from the research conducted by Pereira et al. (2003) and Buck et al. (2011). The total silicon (both elemental and SiO_2) and the amount of plant-available silicon from several sources of silicon are summarized in Table 2.2. The other sources that are not listed in Table 2.2 include mill furnace ashes, crushed basalt, cement, wood biochar and volcanic cinders (Elawad and Green 1979; Savant et al. 1999; Gu et al. 2011; Varela-Milla et al. 2013).

Because of the added value of plant-based silicon sources to overall soil quality, the silicon-rich materials from plant biomass as potential sources of bioavailable silicon were evaluated. The application of biochar improved the soil chemical properties (e.g., the pH and cation exchange capacity, among others) and the soil physical properties, such as water-holding capacity and aggregation (Glaser et al. 2002; Chan et al. 2007). Rice husks are a major waste that is generated by the rice mills, and the carbonized rice husk has been used as an on-farm source of silicon in rice production systems (Sistani et al. 1997; Hossain et al. 2001). The application of biochar (the product of plant biomass pyrolysis) from *Miscanthus* not only increased both soil carbon sequestration and fertility, but also increased the bioavailable silicon that was extracted by $CaCl_2$ solution (Houben et al. 2014). Among the biochars produced from three plant-derived feedstocks (coffee husk, woody material, and *Miscanthus*), the biochar from the *Miscanthus* had the highest release rate of bioavailable silicon at 25.8 mg kg^{-1} ln $(min)^{-1}$. According to Ma and Takahashi (2002), rice straw has been widely used as source of silicon primarily because of the long-term effect (40 years) of rice straw on the plant-available silicon concentrations in soil. The silicon in the rice straw is not fully available in the short-term, but the amount of silicon that becomes plant-available in the long-term could exceed 70 % of the amount applied.

The silicon-rich materials from industrial wastes and plant biomass are applied in large amounts. Because most of these materials are also good liming agents, the

Table 2.2 Total and soluble silicon content of different silicon fertilizer sources

Source	Silicon content: Total Si, %	Silicon content: Soluble Si[a], %	Chemical composition	References
Wollastonite	24.2	3.6	$CaSiO_3$	Sebastian et al. 2013
	24.2	6.5	$CaSiO_3$	Haynes et al. 2013
$MgSiO_3$ (Talc)	28.5	0.1	$MgSiO_3$	Sebastian et al. 2013
Silica gel	46.7	5.8	Not known	Sebastian et al. 2013
K_2SiO_3-liquid	9.7	7.6	K_2SiO_3	Sebastian et al. 2013
$NaSiO_3$-liquid	5.6	–	Na_2SiO_3	Abed-Ashtiani et al. 2012
Silicic acid	36.0	6.4	–	Sebastian et al. 2013
Silica blend (monocal or with $FeSO_4$, NH_4NO_3, KCl)	12.1	1.8	$CaSiO_3$(mainly)	Sebastian et al. 2013
$CaSiO_3$/$MgSiO_3$ blend	12.0	2.2	$CaSiO_3$/$MgSiO_3$	Sebastian et al. 2013
Industrial by-product				
Iron/steel slag	5.4	0.46	$CaSiO_3$	Haynes et al., 2013
Electric furnace slag	21.1 20.3	14.8[b] 0.5	$CaSiO_3$/$MgSiO_3$	Gascho and Korndorfer 1998 Sebastian et al. 2013
Blast furnace slag	17.3	1.7	$CaSiO_3$/$MgSiO_3$	Haynes et al. 2013
Processing mud	6.8	0.04	–	Haynes et al. 2013
Fly ash	29.1 23.0	0.03 0.01	–	Haynes et al. 2013 Raghupathy 1993
Plant material-based silica				
Miscanthus biochar	38.3	–	SiO_2	Houben et al. 2014
Rice hull fresh	7–9.2	–	SiO_2	Sun and Gong 2001
Rich hull ash	>28.0	–	SiO_2	Kalapathy et al. 2002

[a]5-day Na_2CO_3-NH_4NO_3 Soluble Silicon Extraction Method (SLV 5-day)
[b]2 % Citric acid procedure

pH values of the soils that receive these materials commonly increase substantially (Tubaña et al. 2012a; Haynes et al. 2013). Using a liquid silicon formulation has advantages in the ease of application at manageable rates when compared with the sources of solid silicon. Both potassium and sodium silicate solutions are used as either a foliar supplement or a soil drench (Menzies et al. 1992; Bélanger et al. 1995; Kanto et al. 2006; Rodrigues et al. 2009; Kamenidou et al. 2010). In greenhouses with hydroponic crop production systems, the liquid silicon formulation is added to the recirculating nutrient solutions (Adatia and Besford 1986). Additionally, several studies showed positive crop responses to a foliar silicon spray. For example, the rice grain yield increased after a foliar application of soluble silicic acid (Prakash et al. 2011).

Measuring Silicon Concentration

The molybdenum blue colorimetry is commonly used to quantify the silicon concentrations in water and extracted/digested samples (Hallmark et al. 1982). The monosilicic acid is the only form of silicon that is molybdate-reactive, and the other forms of silicon (e.g., polysilicic acid) have little to no effect on the formation of the silicon-molybdate complex. This complex forms an intense blue color in the solution, which increases in intensity with an increase in the concentration of the H_4SiO_4 (Hallmark et al. 1982; Sparkman 2006). Although the concentration of silicon is also measured with the inductively coupled plasma-optical emission spectrometry (ICP-OES), notably, this analysis measures all the forms of silicon in solution, including polysilicic acid, which is not plant-available. The measurement of all forms may confuse the interpretation of the results when the silicon is analyzed in soil extracts because large amounts of polysilicic acid in the soil solution may lead to an overestimation of the plant-available silicon. However, to quantify the total silicon content in plant samples, the ICP-OES analysis may be suitable because for the molybdenum blue colorimetry, fluoride ions must be in the plant digest to facilitate the complete ionization of the polysilicic acids (Iler 1955), which eventually optimizes and stabilizes the absorbance readings (Kraska and Breitenbeck 2010). Moreover, because the molybdenum blue colorimetry is highly sensitive, a large dilution of the sample extract is required, which may magnify any errors of measurement (van der Vorm 1987).

Based on the vast amount of literature, many researchers in general have focused on the standardization of the procedures to extract the different fractions of silicon from the soil. To date, although many procedures have been established and modified for different soil types, no universal method has been accepted as the standard. The methods for plant tissue digestion have also undergone multiple modifications, primarily to simplify the method and to improve the precision.

1. Methods for Extraction of Different Silicon Fractions from Soil In the past 50 years, many procedures were identified and used to extract the different forms of silicon from the soil (Hashimoto and Jackson 1960; Beckwith and Reeve 1963, 1964; Schachtschabel and Heinemann 1967). Sauer et al. (2006) reviewed the various methodologies that are used not only to quantify plant-available silicon, but also to extract silicon from amorphous silica and allophane in soils and sediments. Because the solubility of the amorphous silica markedly increases at higher pH values (Iler 1979), the majority of the extraction procedures use alkaline solutions to quantify the silicon bound in this fraction (Table 2.3).

Table 2.3 Extraction procedures used for determining silicon in the solid phase

Solution	Procedure	Silicon fractions	References
NaOH	0.5 M NaOH; 1 g soil in 50 mL solution; 4 h boiling	Amorphous (biogenic and minerogenic)	Foster 1953
	0.5 NaOH; 2.5 min boiling	Amorphous and oxides	Hashimoto and Jackson 1960
	0.5 M NaOH; light fraction of coarse silt (20–50 μm), filter content in 15 mL solution; 16 h at 150 °C	Amorphous (biogenic)	Herbauts et al. 1994
	0.5 M NaOH; 1 g coarse silt (20–50 μm) sample in 100 mL solution; 20 boiling	Amorphous (biogenic)	Jones 1969
KOH + HCl	2.5 min boiling in 0.5 M KOH solution followed by centrifugation and 1 h shaking with 6 M HCl	Amorphous	McKeyes et al. 1974; Karathanasis 1989
Na_2CO_3	0.5 M Na_2CO_3; sequential extraction; 100 mg clay in 80 mL cold solution for 16 H shaking followed by 2 h boiling; repeat extraction until silicon content is low and constant	Amorphous	Follett et al. 1965
	0.5 M Na_2CO_3; 1 g sample in 25 mL solution; agitate for 10 min at 80 °C, repeat extraction until silicon content is low and constant	Amorphous	Arnseth and Turner 1988
	0.5 M Na_2CO_3; 2 g sample in 50 mL solution; 16 h shaking at room temperature	Amorphous	Breuer 1994; Breuer and Herrmann 1999
NaOH + Na_2CO_3	2 % Na_2CO_3 digestion of iron oxides-pre-extracted samples at 90 °C for 15 min then treated with 0.5 N NaOH and heated at 90 °C for 15 min	Amorphous	Wada and Greenland, 1970
Tiron	0.1 M Tiron (4, 5-dihydroxy-1, 3-benzene-disulfonic acid [disodium salt]) (pH 10.5); 25 mg sample in 30 mL solution; 1 h at 80 °C	Amorphous	Biermans and Baert 1977; Kodama and Ross 1991

A wet chemical dissolution process that uses the strong base NaOH is a standard technique that was developed in 1950s to analyze the amorphous silicon in soils (Foster 1953). Although this procedure also dissolves biogenic and minerogenic silica, the amorphous silica that is bound in the sesquioxides remains intact even under prolonged exposure to a high temperature and an alkaline solution. Nevertheless, a tendency to overestimate the silicon content of soils from this fraction occurs because the silicate minerals partially dissolve using this method and eventually release silicon (Wada and Greenland 1970). This tendency for overestimation has prompted modifications to the length of time for which samples are exposed to a boiling temperature and to the composition of the solutions to (1) ensure that the silicon measured in the extracts is from the dissolved amorphous silicon and (2) effectively remove the amorphous silica from soils (Hashimoto and Jackson 1960; McKeyes et al. 1974; Karathanasis 1989). This standard technique was also modified to specifically quantify the biogenic silica content of soils (Jones 1969; Herbauts et al. 1994), which generally involves a wet chemical dissolution using the NaOH on only the light fraction of the coarse silt (20–50 μm).

Follett and his colleagues (1965) proposed a sequential extraction procedure to quantify the silicon from the graded clay fraction of the soil in which soil samples were subjected to cold (16 h shaking) and hot (2 h boiling) extraction steps with a 5 % Na_2CO_3 solution. The entire extraction procedure was repeated until low and constant levels of silicon were measured in the extracts. In modifications of the procedure, Arnseth and Turner (1988) reduced the shaking time to 10 min and removed the cold extraction step, whereas Breuer (1994) and Breuer and Herrmann (1999) maintained the 16-h shaking time but removed the cold and the hot extraction steps. Nevertheless, Sauer et al. (2006) noted that the silicon fractions dissolved by the modified methods were assumed to be similar to those extracted by the original sequential extraction procedure of Follett et al. (1965)

In 1970, a procedure was established that combined the NaOH and the Na_2CO_3 solutions in the extraction to address the limitations of the NaOH-sequential cold and hot extraction procedure (Wada and Greenland 1970). When subjected to the cold and hot extraction with NaOH, varying amounts of silicic acid were released from pure clay minerals. Thus, Follett et al. (1965) assumed that the silicic acid originated from the amorphous materials, and therefore, the amount and the type of material found in the soil clay (from completely disordered to well crystallized material) that was dissolved was dependent on the type of the solution. However, Wada and Greenland (1970) indicated that initial mineral composition was the predominant influence on this reaction. Later, Krausse et al. (1983) showed that the extent of the mineral dissolution was also dependent on the digestion time, temperature, pH, concentration, and volume of the reagent.

For soil nutrient management, the abundance of silicon in the soil is interpreted differently. Because the agronomic value of silicon fertilization is well recognized in production agriculture, the research interest shifted in recent years, and many methodologies were established to determine the plant-available silicon (Datnoff et al. 2001). The most important fraction of the silicon that is subject to interpretation is the form available for plant uptake, because the amount of plant-available silicon determines whether silicon fertilization is required. The plant-available silicon is presumably composed of silicic acid, both in the liquid (in the soil solution) and in the adsorbed phases (to the soil particles). The suitable solutions identified to extract plant-available silicon include water, $CaCl_2$, acetate, acetic acid, phosphate, H_2SO_3, H_2SO_4, and citrate (Table 2.4). These solutions also extract the desorbed silicic acid, with the H_2SO_3, H_2SO_4, and citrate as the most effective solutions.

The procedures summarized in Table 2.4 experienced a series of modifications, most of which generally resulted in a shorter extraction time. As reported by McKeague and Cline (1963), a prolonged shaking time, even with water only, can

Table 2.4 Extraction procedures used for determining soluble and adsorbed silicon in soil

Solution	Procedure	Silicon fractions	References
H_2O	10 g in 50 mL + 0.1 % NaN_3 to reduce biological activity; incubate 21 days at room temperature with manual shaking 2 times a day	Water-soluble	Schachtschabel and Heinemann 1967
	10 g in 100 mL; 4 h shaking	Water-soluble	Fox et al. 1967; Khalid et al. 1978
	10 g in 60 mL; incubate at 40 °C for 2 weeks	Water-soluble	Nonaka and Takahashi 1988, 1990
	10 g in 100 mL; 1 h shaking	Water-soluble	Korndörfer et al. 1999
$CaCl_2$	0.01 M $CaCl_2$; 1 g sample in 20 mL solution; 16 h shaking	Liquid phase; readily available	Haysom and Chapman 1975
	0.01 M $CaCl_2$; 10 g sample in 100 mL solution; 1 h shaking	Liquid phase; readily available	Korndörfer et al. 1999
Na acetate + acetic acid	0.18 N Na acetate + 0.87 M acetic acid, adjusted to pH 4; 10 g sample in 100 mL solution; 5 h occasional shaking at 40 °C	Soluble and some exchangeable	Imaizumi and Yoshida 1958
	0.18 N Na acetate + 0.87 M acetic acid, adjusted to pH 4; 10 g sample in 100 mL solution; 1 h shaking	Soluble and some exchangeable	Korndörfer et al. 1999

(continued)

Table 2.4 (continued)

Solution	Procedure	Silicon fractions	References
NH_4 acetate	5 % (0.5 M) NH_4 acetate, adjusted to pH 4.5–4.8 with 0.1 M acetic acid; 1 g sample in 20 mL solution; 1 h shaking	Soluble and some exchangeable	Ayres 1966; Cheong and Halais 1970
	5 % (0.5 M) NH_4 acetate, adjusted to pH 4.8 with 0.1 M acetic acid; 1 g sample in 10 mL solution; 1 h shaking	Soluble and some exchangeable	Korndörfer et al. 1999
Acetic acid	0.5 M acetic acid; 1 g sample in 10 mL solution; 1 h shaking with 12 h resting	Soluble and some exchangeable	Snyder 1991
	0.5 M acetic acid; 1 g sample in 10 mL solution; 1 h shaking	Soluble and some exchangeable	Korndörfer et al. 1999
	0.5 M acetic acid; 10 g sample in 25 mL solution; overnight resting followed by 2 h shaking	Soluble and some exchangeable	Snyder 2001
Phosphate acetate	0.016 M P as Ca $(H_2PO_4)_2$ dissolved in 0.1 M NH_4 acetate, adjusted to pH 3.5 with 0.1 M acetic acid; 1 g sample in 10 mL solution; 4 h shaking	Soluble and some exchangeable	Fox et al. 1967
	0.0016 M P as $Ca(H_2PO_4)_2$ dissolve in 0.1 M acetic acid adjusted to pH 3.5; 1 g sample in 10 mL solution; 4 h shaking	Soluble and some exchangeable	Khalid et al. 1978
Citric acid	0.1 M citric acid; 1 g sample in 50 mL solution; 2 h shaking, resting overnight then 1 h shaking	Soluble, exchangeable, and adsorbed	Acquaye and Tinsley 1964
Na citrate + $NaHCO_3$	80 % 0.3 M Na citrate and 20 % 1 M $NaHCO_3$; 2 g sample in 50 mL solution; 5 min at 80 °C	Soluble, exchangeable, and adsorbed to sesquioxide surfaces	Breuer 1994

(continued)

Table 2.4 (continued)

Solution	Procedure	Silicon fractions	References
NH_4 citrate	1 M NH_4 citrate; 10 g sample in 25 mL solution; 80 h shaking	Soluble, exchangeable, and adsorbed	Sauer and Burghardt 2000, 2006
$H_2SO_3+(NH_4)_2SO_4$	0.02 N H_2SO_3 containing 0.02 M $(NH_4)_2SO_4$; 1 g sample in 100 mL solution; 30 min shaking	Soluble, exchangeable, and adsorbed	Fox et al. 1967
H_2SO_4	0.005 M H_2SO_4; 1 g sample in 200 mL solution; 16 h shaking	Soluble, exchangeable, and adsorbed	Hurney 1973

increase the amount of silicon extracted from a soil because of the abrasion. By contrast, a prolonged shaking time results in equilibration between the soil and the solution (Schachtschabel and Heinemann 1967; Nonaka and Takahashi 1988, 1990); however, the time required for the completion of the procedure is too long, and therefore, the adoption of this approach in commercial soil testing laboratories will be limited. With water, the least amount of soluble silicon is extracted in soils, whereas the silicon extracted with $CaCl_2$ is the most easily removed of the soluble fractions (Berthelsen et al. 2001). Haysom and Chapman (1975) reported a high correlation between the silicon extracted with distilled water and that extracted with a 0.01 M $CaCl_2$ solution from the acidic soils of northern Queensland. Mengel and Kirkby (2001) found that the amount of soluble silicon extracted with both distilled water and the $CaCl_2$ solution was primarily H_4SiO_4, which was present at pH values from 2 to 9 and was in equilibrium with the amorphous silica. Nevertheless, the amount of silicon extracted with the $CaCl_2$ solution obtained the highest correlation with the sugar yield ($r^2=0.82$) compared with the silicon extracted with the 0.5 M NH_4 acetate and the 0.005 M H_2SO_4 (Haysom and Chapman 1975).

The soil silicon extracted with the acetic acid/acetate-based solutions is the soluble silicon and some of the exchangeable silicon, primarily the silicon from exchange sites. Nonaka and Takahashi (1990) reported that the amount of silicon extracted by the acetate solution overestimated the plant-available silicon for soils that were previously amended with calcium silicate. Moreover, these authors found that not all the silicon extracted from the calcium silicate was plant-available. Snyder (2001) noted that phosphate buffer used as an extractant did not overestimate the plant-available silicon in soils with silicates applied, likely because the phosphate (anion) only displaced the adsorbed silicic acid rather than dissolving the residual calcium silicate.

The amount of silicon extracted by citric acid, citrate-based solutions and diluted H_2SO_4 is generally higher than the amount extracted with acetate-based solutions. This result was attributed to the silicon contributed by the adsorbed fractions (both particles and hydroxides). Beckwith and Reeve (1964) also noted that the citrate ions not only competed for the sorption sites for silicic acid but also formed complexes with metal ions that are known to bind silicic acid. The assumption of Breuer (1994) that the Na citrate + $NaHCO_3$ solutions extracted the silicon fraction that was specifically adsorbed to the sesquioxide surfaces was supported by the close correlation between the silicon extracted with this method and the silicon content in dithionite extracts (Mehra and Jackson 1960). Although the Na citrate + $NaHCO_3$ solutions extracted only 17 % of the amount of silicon found in the dithionite extracts, a strong correlation ($r^2 = 0.81$) between the silicon that was extracted by these two methods was observed. The dithionite solution effectively dissolves pedogenic sesquioxide; thus, the silicon quantified in the extract originates not only from the surface, but also as the silicon bound inside of the sesquioxides. Similar to phosphate, the sulfate-based solutions were noted to effectively extract silicon from a wider array of soil types than the acetate-based solutions (Fox et al. 1967). The acidity of the citric and sulfate-based solutions combined with a long shaking time (Hurney 1973) chemically and mechanically abraded the silicon from the silicates and the clay minerals, which resulted in an overestimation of the plant-available silicon. The actual amount of readily soluble silicon in the soil regardless of the origin (biogenic or pedogenic) is quantified with alkaline dissolution (Sauer et al. 2006; Saccone et al. 2007; Cornelis et al. 2011).

The amount of silicon that is extracted is different among these procedures, presumably, because the silicon extracted did not originate from the identical fractions. This problem poses a complication for the determination of the silicon fertilizer requirement, because the determination will be based on the choice of the extractant. The assumption is that these solutions all extract the dissolved plant-available silicon. Fox et al. (1967) used $Ca(H_2PO_4)_2$, H_2SO_4, and acetic acid to extract silicon from the soils of Hawaii with different mineral compositions. Based on the results, water consistently extracted the least amount of silicon in all the soils, and in the soils dominated by montmorillonite, kaolinite, goethite and gibbsite, the $Ca(H_2PO_4)_2$ extracted the most silicon. With the exception of the desert soils, the rest of the soils from the volcanic ashes that were dominated by allophane had the highest amount of silicon extracted with the H_2SO_4 as the extractant. The amount of silicon extracted with the acetic acid solution was between the amounts extracted with water and with $Ca(H_2PO_4)_2$ or H_2SO_4. The comparisons of Berthelsen et al. (2001) for different extractants revealed similar results; the solutions that contained diluted H_2SO_4 and citric acid extracted 12- and 16-foldmore silicon than the $CaCl_2$solution, respectively. For calcareous soils, the acidic extractants (e.g., acetic acid and sulfuric acid) tended to remove the greatest amount of silicon, which originated from the highly acid-soluble calcium silicates. However, the silicon of this form was not easily absorbed by plants (Xu et al. 2001). Large amounts of silicon were removed from the acidic volcanic soils of northern Queensland using 0.005 M H_2SO_4 because of the ability to dissolve the sesquioxide compounds that contained the adsorbed silicon Hayson and Chapman (1975).

Tubaña et al. (2012b) and Babu et al. (2013) also demonstrated that the amounts of silicon extracted were variable using different procedures for soils collected from the Midwest and the southern USA. Tubaña et al. (2012b) showed that 0.1 M citric

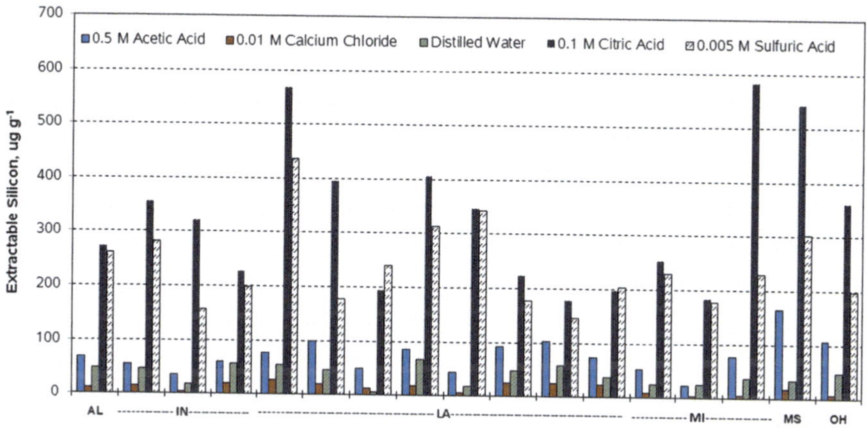

Fig. 2.4 Silicon concentration extracted from soils collected from Alabama, Indiana, Louisiana, Michigan, Mississippi, and Ohio using different extraction procedures (Tubaña et al. 2012b)

acid consistently extracted the largest amount of silicon from soils collected from selected states in the US (Fig. 2.4). Babu et al. (2013) noted that the amount of extractable silicon was in the order of citric acid > acetic acid (24 h rest + 2 h shaking > 1 h shaking) > sodium acetate > ammonium acetate > calcium chloride > water. Similarly, the 0.025 M citric acid solution also extracted higher quantities of silicon in calcareous soils than the Na acetate–acetic acid (pH 4) and the 0.19 M Na_2CO_3–0.5 M $NaHCO_3$ (pH 8.5) solutions (Xu et al. 2001).

Lima Rodrigues et al. (2003) correlated the amounts of silicon extracted by acetic acid (Snyder 2001), acetate/acetic acid (Imaizumi and Yoshida 1958), and $CaCl_2$ (Haysom and Chapman 1975) solutions from soils collected from 31 countries. These soils collectively represented 137 mineral soils, primarily as Oxisols, Ultisols, and coarse-textured soils. The relationship between the silicon extracted with acetic acid and acetate/acetic acid ($r^2=0.59$) was relatively stronger than that between the silicon extracted with acetic acid and $CaCl_2$ ($r^2=0.53$). Babu et al. (2013) obtained similar results for the relationship between the silicon extracted with acetic acid and sodium acetate ($r^2=0.56$) from the soils (~130 samples) of Louisiana that were farmed for different field crops.

Korndörfer et al. (1999) standardized the soil to solution ratio (1:10) and the shaking time (1 h) before filtration for the water, acetic acid, $CaCl_2$, and Na acetate + acetic acid extraction procedures (Table 2.4). The authors used these procedures to determine the silicon contents of four soil types from Brazil that were treated with five levels of wollastonite and were grown with upland rice. For the extraction procedures, the amount of silicon extracted from the soil and the silicon content of the rice had strong correlations (r^2 values > 0.69). Barbosa-Filho et al. (2001) also evaluated a similar set of methods using the predominantly organic soils of the Everglades agricultural areas in southern Florida. In general, the soil silicon values obtained from the different methods were correlated with the straw and the panicle silicon content. However, the soil silicon extracted with 0.5 M acetic acid obtained the highest correlation with the silicon contents of the rice straw (r^2=0.90) and the panicle (r^2=0.84).

2. Soil Silicon Critical Levels To interpret soil tests and to determine fertilization guidelines for a nutrient requires knowledge of the critical level in the soil. The critical nutrient level is the point in a crop response curve that corresponds to the level of a plant-available nutrient that generates the maximum yield. Above the critical nutrient level, fertilization of the crop is in excess, whereas at levels below this point, a crop response to higher levels of fertilization is likely. To date, the published critical silicon levels varied with soil type, crops, and soil testing procedure. The critical silicon levels that were established using the different extraction procedures to determine plant-available silicon are summarized in Table 2.5. Lima Rodrigues et al. (2003) observed that the different extraction procedures would predict different levels of silicon deficiency in soil and therefore, different resultant silicon requirements for a crop. Using the published critical soil test silicon values for the Na acetate buffer, acetic acid and $CaCl_2$ extraction procedures, the authors created a subset of 137 mineral soils from 31 countries. The results of the tests on the subset of soils were at or below the published critical silicon levels for each of the extraction procedures. The silicon extracted with the Na acetate buffer correlated well with the acetic acid (r^2=0.71), but not with the $CaCl_2$ (r^2=0.33).

Based on the calibration tests conducted by Xu et al. (2001) that involved 17 field trials, the sodium acetate + acetic acid solution of Imaizumi and Yoshida (1958) was the optimal extraction method to assess plant-available silicon in calcareous soils. Using wheat biomass as a response variable, the authors established the critical level at 80 mg silicon kg^{-1} (171 mg SiO_2 kg^{-1}). These soils were classified as Inceptisols with a high soil pH that ranged from 7.40 to 8.25 and $CaCO_3$ concentrations that ranged from 26.5 to 52.6 g kg^{-1}. The acetic acid (0.5 M) extraction procedure was suitable for the organic and mineral soils in south Florida that were characterized by low clay, Al, and Fe contents (Korndörfer et al. 2001). The established critical silicon level for these soils was 19 mg kg^{-1} (Table 2.5). Korndörfer et al. (2001) categorized the soil silicon test values below this critical level such that >24, 6–24, and <6 mg silicon kg^{-1} soil were interpreted as high, medium and low soil silicon test values, respectively, and should be fertilized with 0, 1120, and 1500 kg silicon ha^{-1}, respectively. The calcium chloride (0.01 M) extraction procedure was developed in Australia as an alternative to the extraction method of distilled water; the latter method has problems with the interference by dispersed clay fractions in a water suspension (Haysom and Chapman 1975). These authors established the soil silicon critical level based on the extraction with calcium chloride solution at 20 mg kg^{-1} for sugarcane. McCray et al. (2011) reported that the soil silicon critical level based on an acetic acid extraction of the soils used to grow sugarcane in Florida was 32 g m^{-3}.

3. Methods to Extract Silicon from Plant Tissue Samples The silicon from plant tissue samples can be extracted using a gravimetric method, a hydrofluoric acid solubilization, an autoclave-induced digestion with a strong NaOH solution or a microwave digestion assisted with nitric and hydrofluoric acids (Yoshida et al. 1976;

Table 2.5 Critical silicon levels established in different soils using different extraction procedures for different crops

Solutions	Critical levels mg Si kg^{-1}	Soil types/ orders	Crops	References
Acetic acid	19	Histosols	Rice	Snyder 1991; Korndörfer et al. 2001
	54	Acid/Ultisol	Rice	Narayanaswamy and Prakash 2009
	32[a]	Histosol	Sugarcane	McCray et al. 2011
Acetic acid w/ 24 h rest	87	Acid/Ultisol	Rice	Narayanaswamy and Prakash 2009
Acetate – buffer	33 28	Tropical soils	Rice	Kawaguchi 1966 Takijima et al. 1970
Na acetate – acetic acid	80	Calcareous	Wheat	Xu et al. 2001
	71–181	Calcareous	Rice, Wheat	Liang et al. 1994
	38	Acid and neutral	Rice	Takijima et al. 1970
	60	Acid and neutral	Rice	Imaizumi and Yoshida 1958
	38–60	Acid and neutral	Rice	He 1980
	60	Acid and neutral	Rice	Lian 1976
Na acetate	85	Acid/Ultisol	Rice	Narayanaswamy and Prakash 2009
NH_4 acetate	32	Acid/Ultisol	Rice	Narayanaswamy and Prakash 2009
H_2O – 1 h shaking	14	Acid/Ultisol	Rice	Narayanaswamy and Prakash 2009
H_2O – 4 h shaking	30	Acid/Ultisol	Rice	Narayanaswamy and Prakash 2009
$CaCl_2$	43	Acid/Ultisol	Rice	Narayanaswamy and Prakash 2009
	20	Acid	Sugarcane	Haysom and Chapman 1975
Citric acid	185	Acid/Ultisol	Rice	Narayanaswamy and Prakash 2009
H_2SO_4	207	Acid/Ultisol	Rice	Narayanaswamy and Prakash 2009

[a]Critical level of Si expressed as g/m^3

NovozamskyI et al. 1984; Elliot and Snyder 1991; Feng et al. 1999). The silicon in these extracts is analyzed colorimetrically or by ICP-OES. The standardization of the procedures in silicon plant tissue testing has not encountered as many challenges as the standardization of the soil silicon testing. The modifications of the procedures were to address safety (e.g., the nitric acid-hydrofluoric acid digestion), the availability of instruments, and the time-consuming and difficult-to-perform procedures (Fox et al. 1969; Elliot and Snyder 1991; Ostatek-Boczynski and Haysom 2003). The gravimetric method, which was established in 1960 by Horwitz, is time consuming and requires platinum ware at each stage of the extraction. The bomb technique proposed by NovozamskyI et al. (1984) uses a mixture of very reactive and hazardous chemicals (hydrochloric, nitric, and hydrofluoric acids) during autoclaving. The procedures were also modified to ensure accurate and reproducible results. Whereas the widely used autoclave-induced digestion method of Elliot and Snyder (1991) is relatively rapid, inexpensive and specialized instrumentation is not required (Bell and Simmons 1997), others have reported that the results are highly variable and tend to underestimate the silicon content of the plant (Taber et al. 2002; Haysom and Ostatek-Boczynski, 2006). The underestimation of the silicon values in plant tissue samples is attributed to the vigorous foaming that occurs when the H_2O_2 and the NaOH are combined in the sample tube, which deposits samples on the upper tube wall. During autoclaving, the sample particles on the upper tube wall are not well digested. Before the addition of the hydrogen peroxide, the addition of five drops of octyl-alcohol was incorporated into the method (Modified Autoclave Digestion–MAD) to eliminate the excessive foaming. The MAD procedure was later simplified to use an oven instead of the autoclave during digestion (Kraska and Breitenbeck 2010). To ensure that the color development is stable during the colorimetric procedure, the addition of 1 mL of 5 mM ammonium fluoride was also introduced to the procedure, now the Oven-Induced Digestion (OID). The ammonium fluoride ions facilitate the complete ionization of the polysilicic acid in the plant digests, which provides for more stable absorbance readings.

Among the published studies on which plant part should be analyzed for the concentration of silicon, there is relatively good agreement. For example, for the most practical testing procedure, the straw of rice plants at harvest was used as the sample material. Park et al. (1964) used the rice flag leaf as the sample material for silicon content determination, which they also used as an index of the available silicon in soil. To attain high levels of accuracy and sensitivity, low coefficients of variation and practical convenience, Winslow (1995) proposed the use of rice hull as the sample material for silicon content determinations in rice.

4. Critical Silicon Concentration in Plant Tissue Samples The plant silicon content is an accepted parameter for the routine monitoring of the silicon status in crops. Currently, only a few published critical silicon levels in plant tissue are available, and these were published primarily for rice and sugarcane. The critical silicon content in rice straw was established at 37 g kg^{-1} by Nair and Aieyer (1968) and Takijima et al. (1970). The critical silicon level established by Snyder et al. (1986) for rice straw was 30 g kg^{-1}; a value that was closer to 37 g kg^{-1} than the level reported by DeDatta in 1981 at 5 g kg^{-1}. Using the Y-leaf of rice, Dobermann and Fairhurst (2000) reported a similar critical level to that of De Datta (1981). Lian (1976) reported that the critical silicon levels for rice straw as the sample material were 51, 47, and 42 g kg^{-1} for Japan, Korea, and Taiwan, respectively. Korndörfer et al. (2001) established a range of critical levels using straw as the sample material (from 17 to 34 g kg^{-1}) for rice grown in Florida soils; these values are lower than the critical levels reported by Lian (1976) for other rice producing countries. Narayanaswamy and Prakash (2009) established a critical level at 29 g kg^{-1} for straw and at 12 g kg^{-1} for grain for rice grown in southern India.

Only a few studies were conducted to determine the critical silicon level in sugarcane. A narrow range of critical silicon levels in sugarcane leaf was established at 10 g kg^{-1} (Anderson and Bowen 1990), 5.5 g kg^{-1} (Bethelsen et al. 2003), and 5 g kg^{-1} (McCray and Mylavarapu 2010). The critical silicon level that was established many decades ago by Halais (1967), who used the sixth leaf sheath, was the highest (12.5 g kg^{-1}) among the published critical silicon levels.

To attain satisfactory yields, the silicon content in a plant should be above the reported critical silicon level (Snyder et al. 1986). The critical silicon levels currently reported are very specific not only to the crop species but also to the location and the sample material used, which underscores the necessity to establish site-specific plant-silicon content interpretations.

Conclusions

The benefits of silicon to a wide variety of crops are well-documented and strongly demonstrate the value of silicon fertilization in agriculture. Agricultural areas under intensive cropping systems, especially those with soils inherently low in soluble silicon, are amended with silicon-rich materials to ensure plant productivity. In fact, in some parts of the world silicon fertilization is an accepted agronomic practice. While the development and standardization of different procedures to extract and quantify different silicon fractions in soils is considered significant progress in silicon research specifically and the realm of soil science more generally, their applications in soil fertility and nutrient management have been very limited. The development of soil silicon interpretation test and fertilization guidelines in crop production require the establishment of critical soil silicon levels and robust, high-precision soil testing procedures suitable for a wide array of soil types. Thus far, a few extraction procedures (e.g., 0.5 M acetic acid and 0.01 M $CaCl_2$) have been identified and are rigorously employed in calibration/correlation research in many parts of the world, including the US and Brazil. Initial critical soil-based silicon levels using these procedures have been reported but appear to require further refinement. No elaborate soil interpretation test has been derived from these calibration/correlation studies. A soil interpretation test can be used as a tool to determine whether silicon fertilization is needed or not, but it does not provide the concentration of silicon required to raise plant-available silicon to a desired level, nor does it indicate the probability that the crop in question will respond to and benefit from silicon fertilization. The availability of high-precision method(s) for quantifying plant-available silicon in silicon fertilizer is equally as important as an established, well-refined soil silicon interpretation test in providing effective silicon recommendations. One remarkable achievement in silicon research was the development and recognition of the 5-days Na_2CO_3-NH_4NO_3 method for extracting plant-available silicon from solid fertilizer. This method is currently being evaluated in terms of it applicability to many silicon-containing fertilizers. Clearly, silicon research has made progress, particularly in those areas that are critical to the development of effective silicon fertilization guidelines. Even so, there are many soil science aspects of silicon that are understudied (e.g., chemical dynamics and soil-plant interaction). It is strongly believed that the outcomes from these future soil science-based research studies on silicon will significantly advance the current established knowledge of silicon in soil and fertilization guidelines for crop production.

References

Abed-Ashtiani F, Kadir J, Selamat A et al (2012) Effect of foliar and root application of silicon against rice blast fungus in MR219 rice variety. Plant Pathol J 28:164–171

Acquaye D, Tinsley J (1964) Soluble silica in soils. In: Hallsworth EG, Crawford DV (eds), Experimental pedology. In: Proceedings of the eleventh Easter School in Agricultural Science, University of Nottingham, 1964, Butterworths, London, pp 126–148

Adatia MH, Besford RT (1986) The effects of silicon on cucumber plants grown in recirculating nutrient solution. Ann Bot 58:343–351

Al-Aghabary K, Zhu Z, Shi QH (2004) Influence of silicon supply on chlorophyll content, chlorophyll fluorescence, and antioxidative enzyme activities in tomato plants under salt stress. J Plant Nutr 27:2101–2115

Allen BL, Hajek BF (1989) Mineral occurrence in soil environments. In: Minerals in soil environments. Soil Science Society of America, Madison, pp 199–279

Allmaras RR, Laird DA, Douglas CL et al (1991) Long-term tillage, residue management and nitrogen fertilizer influences on soluble silica in Haploxerol. Am Soc Agron, Madison, p 323

Anderson DL, Bowen JE (1990) Sugarcane nutrition. Potash & Phosphate Institute, Atlanta, p 31

Aoki Y, Hoshino M, Matsubara T (2007) Silica and testate amoebae in a soil under pine-oak forest. Geoderma 142:29–35

Arnseth RW, Turner RS (1988) Sequential extraction of iron, manganese, aluminium and silicon in soil from two contrasting water sheds. Soil Sci Soc Am J 52:1801–1807

Ayres AS (1966) Calcium silicate slag as a growth stimulant for sugarcane on low-silicon soils. Soil Sci 101:216–227

Babu T, Datnoff LE, Yzenas J et al (2013) Silicon status of Louisiana soils grown to different field crops. ASA-CSSA-SSSA international annual meetings, Tampa, pp 3–6, Nov 2013

Barbosa-Filho MP, Snyder GH, Elliott C et al (2001) Evaluation of soil test procedures for determining rice-available silicon. Commun Soil Sci Plant Anal 32:1779–1792

Basile-Doelsch I, Meunier JD, Parron C (2005) Another continental pool in the terrestrial silicon cycle. Nature 433:399–402

Baylis AD, Gragopoulou C, Davidson KJ et al (1994) Effects of silicon on the toxicity of aluminum to soybean. Commun Soil Sci Plant Anal 25:537–546

Bazilevich NI (1993) The biological productivity of North Eurasian ecosystems. RAS Institute of Geography, Moscow, p 293

Beckwith RS, Reeve R (1963) Studies on soluble silica in soils. I. The sorption of silicic acid by soils and minerals. Aust J Soil Res 1:157–168

Beckwith RS, Reeve R (1964) Studies on soluble silica in soils. II. The release of monosilicic acid from soils. Aust J Soil Res 2:33–45

Bélanger RR, Bowen PA, Ehret DL et al (1995) Soluble silicon its role in crop disease management of greenhouse crops. Plant Dis 79:329–336

Bélanger RR, Benhamou N, Menzies JG (2003) Cytological evidence of an active role of silicon in wheat resistance to powdery mildew (*Blumeria graminis* f.sp. *tritici*). Phytopathology 93:402–412

Bell PF, Simmons TF (1997) Silicon concentrations of biological standards. Soil Sci Soc Am J 61:321–322

Berthelsen S, Noble AD, Garside AL (2001) Silicon research down under: past, present and future. In: Snyder GH, Korndörfer GH, Datnoff LE (eds) Silicon deposition in higher plants. Silicon in agriculture. Elsevier, New York, pp 241–255

Berthelsen S, Noble AD, Kingston G et al (2003) Improving yield and ccs in sugarcane through the application of silicon based amendments. Final report – SRDC project CLW009. Sugar Research and Development Corporation. CSIRO Land and Water PMBAitkenvale, Townsville, p 54

Bharwana SA, Ali S, Farooq MA et al (2013) Alleviation of lead toxicity by silicon is related to elevated photosynthesis, antioxidant enzymes suppressed lead uptake and oxidative stress in cotton. J Bioremed Biodeg 4:187. doi:10.4172/2155-6199.1000187

Biermans V, Baert L (1977) Selective extraction of the amorphous Al, Fe and Si oxides using an alkaline Tiron solution. Clay Miner 12:127–135

Blecker SW, McCulley RL, Chadwick OA et al (2006) Biological cycling of silica across a grassland bioclimosequence. Glob Biogeochem Cycles 20:1–11

Bocharnikova EA, Matichenkov VV, Pinsky DL (1995) The influence of soluble silica acids on behavior of heavy metals in soil and natural waters. In: Proceedings of the world-wide symposium pollution in large cities, Venece/Padova, p 43

Bouzoubaa Z (1991) Etude de rôles de la silice dans les mécanimes de tolérance à la sécheresse chez quelques espèces de grande culture. PhD thesis, Montpellier II University

Breuer J (1994) Hartsetzende Bö den Nordkameruns. PhD thesis, Soil Science Department, Technical University of Munich, Weihenstephan, p 189

Breuer J, Herrmann L (1999) Eignung der Extraktion mit Natriumbikarbonat für die Charakterisierung von bodenbildendenProzessen. Mittlg Dtsch Bodenk Ges 91:1375–1378

Brown TH, Mahler RL (1987) Effects of phosphorus and acidity on levels of silica extracted from a Palouse silt loam. Soil Sci Soc Am J 51:674–677

Brown TH, Mahler RL (1988) Relationships between soluble silica and plow pans in Palouse silt loam soils. Soil Sci 145:317–393

Buck GB, Korndörfer GH, Datnoff LE (2011) Extractors for estimating plant available silicon from potential silicon fertilizer sources. J Plant Nutr 34:272–282

Casey WH, Kinrade SD, Knight CT et al (2003) Aqueous silicate complexes in wheat, *Triticum aestivum* L. Plant Cell Environ 27:51–54

Chadwick OA, Hendricks DM, Nettleton WD (1987) Silica in duric soils, 2. Mineralogy. Soil Sci Soc Am J 51:982–985

Chan KY, van Zwieten BL, Meszaros I et al (2007) Agronomic values of green waste biochars as a soil amendments. Aust J Soil Res 45:437–444

Chen HM, Zheng CR, Tu C et al (2000) Chemical methods and phytoremediation of soil contaminated with heavy metals. Chemosphere 41:229–234

Cheong WY, Halais P (1970) Needs of sugarcane for silicon when growing in highly weathered latosols. Exp Agric 6:99–106

Cocker KM, Evans DE, Hodson MJ (1998) The amelioration of aluminum toxicity by silicon in higher plants: solution chemistry or an in planta mechanism? Physiol Plant 104:608–614

Conley DJ (2002) Terrestrial ecosystems and the global biogeochemical silica cycle. Glob Biogeochem Cycles 16:1121

Conley DJ, Sommer M, Meunier JD et al (2006) Silicon in the terrestrial biogeosphere. In: Ittekot V, Humborg C, Garnier J (eds) Land–ocean nutrient fluxes: silica cycle. SCOPE Series 66:13–28

Cornelis JT, Titeux H, Ranger J et al (2011) Identification and distribution of the readily soluble silicon pool in a temperate forest soil below three distinct tree species. Plant Soil 342:369–378

Cornell RM, Schwertmann U (1996) The iron oxides: structure, properties, reactions, occurence and uses. VCH, Weinheim/New York

Cotterill JV, Watkins RW, Brennon CB et al (2007) Boosting silica levels in wheat leaves reduces grazing by rabbits. Pest Manag Sci 63:247–253

Da Cunha KPV, Nascimento CWA (2009) Silicon effects on metal tolerance and structural changes in maize (*Zea mays* L.) grown on a cadmium and zinc enriched soil. Water Air Soil Pollut 197:323–330

Datnoff LE, Snyder GH, Deren CW (1992) Influence of silicon fertilizer grades on blast and brown spot development and yields of rice. Plant Dis 76:1011–1013

Datnoff LE, Deren CW, Snyder GH (1997) Silicon fertilization for disease management of rice in Florida. Crop Prot 16:525–531

Datnoff LE, Snyder GH, Korndörfer GH (2001) Silicon in agriculture. Elsevier, Dordrecht

De Datta SK (1981) Principles and practices of rice production. Wiley, New York, p 357

De Lima Rodrigues L, Daroub SH, Rice WR et al (2003) Comparison of three soil test methods for estimating plant-available silicon. Commun Soil Sci Plant Anal 34:2059–2071

De Saussure NT (1804) Recherches chimiques sur la végétation. Chez la veuve Nyon Ed., Gauthiers-Villars, Paris, p 112

Delstanche S, Opfergelt S, Cardinal D et al (2009) Silicon isotopic fractionation during adsorption of aqueous monosilicic acid onto iron oxide. Geochim Cosmochim Acta 73:923–934

Deren CW, Datnoff LE, Snyder GH (1992) Variable silicon content of rice cultivars grown on Everglades Histosols. J Plant Nutr 15:2363–2368

Desplanques V, Cary L, Mouret JC et al (2006) Silicon transfers in a rice field in Camargue (France). J Geochem Explor 88:190–193

Dietrich D, Hinke S, Baumann W et al (2003) Silica accumulation in *Triticum aestivum* L. and *Dactylis glomerata* L. Anal Bioanal Chem 376:399–404

Dietzel M (2000) Dissolution of silicates and the stability of polysilicic acid. Geochim Cosmochim Acta 64:3275–3281

Dietzel M (2002) Interaction of polysilicic and monosilicic acid with mineral surfaces. In: Stober I, Bucher K (eds) Water–rock interaction. Kluwer, Dordrecht, pp 207–235

Dobermann A, Fairhurst TH (2000) Rice: nutrient disorders and nutrient management. Potash and Phosphate Inst., Singapore and IRRI, Manila

Doucet FJ, Schneider C, Bones SJ et al (2001) The formation of hydroxyalumino silicates of geochemical and biological significance. Geochim Cosmochim Acta 65:2461–2467

Dove PM (1995) Kinetic and thermodynamic controls on silica reactivity in weathering environments. In: Chemical weathering rates of silicate minerals. Mineralogical Society of America and the Geochemical Society, Rev. Mineral Geochem 31:235–290

Dove PM, Han N, Wallace AF et al (2008) Kinetics of amorphous silica dissolution and the paradox of the silica polymorphs. Proc Natl Acad Sci U S A 105:9903–9908

Drees LR, Wilding LP, Smeck NE et al (1989) Silica in soils: quartz and disorders polymorphs. In: Minerals in soil environments. Soil Science Society of America, Madison, pp 914–974

Elawad SH, Green VE (1979) Silicon and the rice plant environment: a review of recent research. Riso 28:235–253

Elliot CL, Snyder GH (1991) Autoclave-induced digestion for the colorimetric determination of silicon in rice straw. J Agric Food Chem 39:1118–1119

Eneji AE, Inanaga S, Muranaka S et al (2005) Effect of calcium silicate on growth and dry matter yield of *Chloris gayana* and *Sorghum sudanense* under two soil water regimes. Grass Forage Sci 60:393–398

Eneji AE, Inanaga S, Muranaka S et al (2008) Growth and nutrient use in four grasses under drought stress as mediated by silicon fertilizer. J Plant Nutr 31:355–365

Epstein E (1999) Silicon. Annu Rev Plant Physiol Plant Mol Biol 50:641–664

Epstein E (2001) Silicon in plants: facts vs. concepts. In: Datnoff LE, Snyder GH, Korndörfer GH (eds) Silicon in agriculture. Elsevier, Amsterdam, pp 1–15

Epstein E, Bloom A (2005) Mineral nutrition of plants: principles and perspectives, 2nd edn. Sinauer Associates, Sunderland, p 380

Exley C (1998) Silicon in life: a bioinorganic solution to bioorganic essentiality. J Inorg Biochem 69:139–144

Farmer V, Delbos E, Miller JD (2005) The role of phytolith formation and dissolution in controlling concentrations of silica in soil solutions and streams. Geoderma 127:71–79

Fawe A, Menzies JG, Chérif M et al (2001) Silicon and disease resistance in dicotyledons. In: Datnoff LE, Snyder GH, Korndörfer GH (eds) Silicon in agriculture. Elsevier, Amsterdam, pp 159–170

Feng X, Wu S, Wharmby A et al (1999) Microwave digestion of plant and grain standard reference materials in nitric and hydrofluoric acids for multi-elemental determination by inductively coupled plasma mass spectrometry. J Anal At Spectrom 14:939–946

Fernandes ML, Macias F (1987) Variacion estansional de la composicion de la disolucion de suelos de Galicia en relacion con el tipo de horizonte y material original. An Edafol Y Agrobiol 46:53–65

Follett EAC, McHardy WJ, Mitchell BD (1965) Chemical dissolution techniques in the study of soil clays: Part I. Clay Miner 6:23–34

Foster MD (1953) The determination of free silica and free alumina in montmorillonites. Geochim Cosmochim Acta 3:143–154

Fox RL, Silva JA, Younge OR et al (1967) Soil and plant silicon and silicate response by sugarcane. Soil Sci Soc Am Proc 31:775–779

Fox RL, Silva JA, Plucknett DL et al (1969) Soluble and total silicon in sugarcane. Plant Soil 30:81–91

Foy CD (1992) Soil chemical factors limiting plant root growth. Adv Soil Sci 19:97–149

Fraysse F, Pokrovsky OS, Schott J et al (2006) Surface properties, solubility and dissolution kinetics of bamboo Phytoliths. Geochim Cosmochim Acta 70:1939–1951

Fraysse F, Pokrovsky OS, Meunier JD (2010) Experimental study of terrestrial plant litter interaction with aqueous solutions. Geochim Cosmochim Acta 74:70–84

Gascho GJ (2001) Silicon sources for agriculture. In: Datnoff LE, Snyder GH, Korndörfer GH (eds) Silicon in agriculture, Vol 8, Studies in plant science. Elsevier, pp 197–207. doi:10.1016/s0928-3420(01)80016-1

Gascho GJ, Korndörfer GH (1998) Availability of silicon from several sources determined by chemical and biological methods. In: Soil Science Society of America annual meeting, Baltimore, p 308, 18–22 Oct

Gladkova KF (1982) The role of silicon in phosphate plant nutrition. Agrochemistry 2:133

Glaser B, Lehmann J, Zech W (2002) Ameliorating physical and chemical properties of highly weathered soils in the tropics with charcoal – a review. Biol Fertil Soils 35:219–230

Gomes FB, Moraces JC, Santos CD et al (2005) Resistance induction in wheat plants by silicon and aphids. Sci Agric 62:547–551

Gong HJ, Zhu XY, Chen KM et al (2005) Silicon alleviates oxidative damage of wheat plants in pots under drought. Plant Sci 169:313–321

Goussain MM, Prado E, Moraes JC (2005) Effect of silicon applied to wheat plants on the biology and probing behavior of the greenbug *Schizaphis graminum* (Rond.) (Hemiptera: Aphididae). Neotrop Entomol 34:807–813

Gu H, Qui H, Tian T et al (2011) Mitigation effects of silicon rich amendments on heavy metal accumulation in rice (*Oryza sativa* L.) planted on multi-metal contaminated acidic soil. Chemosphere 83:1234–1240

Gunes A, Inal A, Bagci EG (2007) Silicon-mediated changes on some physiological and enzymatic parameters symptomatic of oxidative stress in barley grown in sodic-B toxic soil. J Plant Physiol 164:807–811

Guntzer F, Keller G, Meunier J (2012) Benefits of plant silicon for crops: a review. Agron Sustain Dev 32:201–213

Guo W, Hou YL, Wang SG (2005) Effect of silicate on the growth and arsenate uptake by rice (*Oryza sativa* L.) seedlings in solution culture. Plant Soil 272:173–181

Haak E, Siman G (1992) Field experiments with Oyeslag (Faltorsok med Oyeslag). Report 185, Uppsala

Halais P (1967) Si, Ca, Mn contents of cane leaf sheaths, a reflexion of pedogensis. In: 1966 Report of the Mauritius Sugarcane Industry Research Institute, Mauritius, pp 83–85

Hallmark CT, Wilding LP, Smeck NE (1982) Silicon. In: Page AL, Miller RH, Keeney DR (eds) Methods of soil analysis, Part 2, 2nd edn, Agronomy series No 9. American Society of Agronomy, Madison, pp 263–273

Hammond KE, Evans DE, Hodson MJ (1995) Aluminium/silicon interactions in barley (*Hordeum vulgare* L.) seedlings. Plant Soil 173:89–95

Hansen HCB, Raben-Lange B, Raulund-Rasmussen K et al (1994) Monosilicate adsorption by ferrihydrite andgoethite at pH 3–6. Soil Sci 158:40–46

Hashimoto J, Jackson ML (1960) Rapid dissolution of allophane and kaolinite-halloysite after dehydration. In: Proceedings of the 7th national conference on clays and clay minerals, Washington, DC, pp 102–113

Hattori T, Inanaga S, Tanimoto E et al (2003) Silicon-induced changes in viscoelastic properties of sorghum root cell walls. Plant Cell Physiol 44:743–749

Hattori T, Inanaga S, Araki H et al (2005) Application of silicon enhanced drought tolerance in *Sorghum bicolor*. Physiol Plant 123:459–466

Haynes RJ, Belyaeva ON, Kingston G (2013) Evaluation of industrial wastes as sources of fertilizer silicon using chemical extractions and plant uptake. J Plant Nutr Soil Sci 176:238–248

Haysom MBC, Chapman LS (1975) Some aspects of the calcium silicate trials at Mackay. Proc Aust Sugar Cane Technol 42:117–122

Haysom MB, Ostatek-Boczynski ZA (2006) Rapid wet oxidation procedure for the estimation of silicon in plant tissue. Commun Soil Sci Plant Anal 37:2299–2306

He DY (1980) Silicon in soils and plants. Proc Soil Sci 5:1–10

Heckman JR, Johnston S, Cowgill W (2003) Pumpkin yield and disease response to amending soil with silicon. HortScience 38:552–554

Henriet C, Draye X, Oppitz I et al (2006) Effects of distribution and uptake of silicon in banana (*Musa* spp.) under controlled conditions. Plant Soil 287:359–374

Herbauts J, Dehalu FA, Gruber W (1994) Quantitative determination of plant opal content in soils using a combined method of heavy liquid separation and alkali dissolution. Eur J Soil Sci 45:379–385

Hiemstra T, Barnett MO, van Riemsdijk WH (2007) Interaction of silicic acid with goethite. J Colloid Interf Sci 310:8–17

Hodson MJ, Sangster AG (1988) Observations on the distribution of mineral elements in the leaf of wheat (*Triticum aestivum* L.) with particular reference to silicon. Ann Bot 62:463–471

Hodson MJ, White PJ, Mead A (2005) Phylogenetic variation in the silicon composition of plants. Ann Bot 96:1027–1046

Horst WJ, Fecht M, Naumann A et al (1999) Physiology of manganese toxicity and tolerance in *Vigna unguiculata* (L) Walp. Z Pflanzenernahr Bodenkd 162:263–274

Hossain KA, Horiuchi T, Miyagawa S (2001) Effects of silicate materials on growth and grain yield of rice plants grown in clay loam and sandy loam soils. J Plant Nutr 24:1–13

Houben D, Sonnet P, Cornelis J (2014) Biochar from *Miscanthus*: a potential silicon fertilizer. Plant Soil 374:871–882

Huang PM (1991) Ionic factors affecting the formation of short-rangeordered aluminosilicates. Soil Sci Soc Am J 55:1172–1180

Hunt JW, Dean AP, Webster RE (2008) A novel mechanism by which silica defends grasses against herbivore. Ann Bot 102:653–656

Hurney AP (1973) A progress report on the calcium silicate investigations. Proc Aust Sugar Cane Technol 40:109–113

Iler RK (1955) The colloid chemistry of silica and silicates. Cornell University Press, Ithaca

Iler RK (1979) The chemistry of silica. Wiley, New York, p 621

Imaizumi K, Yoshida S (1958) Edaphological studies on silicon supplying power of paddy soils. Bull Natl Inst Agric Sci B 8:261–304
Inal A, Pilbeam DJ, Gunes A (2009) Silicon increases tolerance to boron toxicity and reduces oxidative damage in barley. J Plant Nutr 32:112–128
Iwasaki K, Maier P, Fecht M et al (2002a) Effects of silicon supply on apoplastic manganese concentrations in leaves and their relation to manganese tolerance in cowpea (*Vigna unguiculata* (L.) Walp.). Plant Soil 238:281–288
Iwasaki K, Maier P, Fecht M et al (2002b) Leaf apoplastic silicon enhances manganese tolerance of cowpea (*Vigna unguiculata*). J Plant Physiol 159:167–173
Janislampi KW (2012) Effect of silicon on plant growth and drought stress tolerance. All graduate theses and dissertations. Paper 1360. http://digitalcommons.usu.edu/etd/1360
Jones RL (1969) Determination of opal in soil by alkali dissolution analysis. Soil Sci Soc Am Proc 33:976–978
Jones LHP, Handreck KA (1963) Effects of iron and aluminumoxides on silica in solution in soils. Nature 198:852–853
Jones LHP, Handreck KA (1965) Studies of silica in the oat plant. III. Uptake of silica from soils by plant. Plant Soil 23:79–96
Jones LHP, Handreck KA (1967) Silica in soils, plants, and animals. Adv Agron 19:107–149
Kalapathy U, Proctor A, Shultz J (2002) An improved method for production of silica from rice hull ass. Bioresour Technol 85:285–289
Kamenidou S, Cavins TJ, Marek S (2010) Silicon supplements affect floricultural quality traits and elemental nutrient concentrations of greenhouse produced gerbera. Sci Hortic 123:390–394
Kanto T, Miyoshi A, Ogawa T et al (2006) Suppressive effect of liquid potassium silicate on powdery mildew of strawberry in soil. J Gen Plant Pathol 72:137–142
Karathanasis AD (1989) Solution chemistry of Fragipans-thermodynamic approach to understanding Fragipan formation. In: Fragipans: their occurrence, classification, and genesis. SSSA Spec Publ 24:113–139
Karathanasis AD (2002) Mineral equilibria in environmental soil systems. In: Soil mineralogy with environmental applications. Soil Science Society of America, Madison, pp 109–151
Kawaguchi K (1966) Tropical paddy soils. Jpn Agric Res Q 1:7–11
Kaya C, Tuna AL, Sonmez O et al (2009) Mitigation effects of silicon on maize plants grown at high zinc. J Plant Nutr 32:1788–1798
Keller WD (1955) The principle of chemical weathering. Columbia Mo, Lucas Bros, p 111
Khalid RA, Silva JA, Fox RL (1978) Residual effects of calcium silicate in tropical soils: II. Biological extraction of residual soil silicon. Soil Sci Soc Am J 42:94–97
Kidd PS, Llugany M, Poschenrieder C (2001) The role of root exudates in aluminum resistance and silicon-induced amelioration of aluminum toxicity in three varieties of maize (*Zea mays* L.). J Exp Bot 52:1339–1352
Kirkham MB (2006) Cadmium in plants on polluted soils: effects of soil factors, hyperaccumulation, and amendments. Geoderma 137:19–32
Knight CTG, Kinrade SD (2001) A primer on the aqueous chemistry of silicon. In: Datnoff LE, Snyder GH, Korndörfer GH (eds) Silicon in agriculture. Elsevier, Amsterdam, pp 57–84
Kodama H, Ross GJ (1991) Tiron dissolution method used to remove and characterize inorganic components in soils. Soil Sci Soc Am J 55:1180–1187
Komisarov ID, Panfilova LA (1987) The method for production of slowly soluble fertilizer. USSR patent 1353767, Patent pending 15 May 84. Published in Bull. Inventions, no 4
Korndörfer GH, Coelho MN, Snyder GH et al (1999) Avaliação de métodos de extração de silício para solos cultivados com arroz de sequeiro. Rev Bras Ci Solo 23:101–106
Korndörfer GH, Snyder GH, Ulloa M et al (2001) Calibration of soil and plant silicon analysis for rice production. J Plant Nutr 24:1071–1084
Kovda VA (1985) Biogeochemistry of soil cover. Nauka Publication, Moscow, pp 159–179

Kraska B, Breitenbeck G (2010) Simple, robust method for quantifying silicon in plant tissue. Comm Soil Sci Plant Anal 41:2075–2085
Krausse GL, Schelske CL, Davis CO (1983) Comparison of three wet-alkaline methods of digestion of biogenic silica in water. Freshwater Biol 13:73–81
Kurtz AC, Derry LA, Chadwick OA (1987) Accretion of Asiandust to Hawaiian soils: isotopic, elemental and mineral mass balances. Geochim Cosmochim Acta 65:1971–1983
Lanning FC (1963) Nature and distribution of silica in strawberry plants. Proc Am Soc Hort Sci 76:349–358
Li Q, Ma C, Shang Q (2007) Effects of silicon on photosynthesis and antioxidant enzymes of maize under drought stress. Yingyong Shengtai Xuebao 18:531–536
Lian S (1976) Silica fertilization of rice. In: The fertility of paddy soils and fertilizer application for rice. Food and Fertilizer Technology Center, Taipei, pp 197–221
Liang YC (1999) Effects of silicon on enzyme activity, and sodium, potassium and calcium concentration in barley under salt stress. Plant Soil 209:217–224
Liang TC, Ding RX (2002) Influence of silicon on microdistribution of mineral ions in roots of salt-stressed barley as associated with salt tolerance in plants. Sci China Ser C 45:298–308
Liang YC, Ma TS, Li FJ et al (1994) Silicon availability and response of rice and wheat to silicon in calcareous soils. Commun Soil Sci Plant Anal 25:2285–2297
Liang YC, Shen QR, Shen ZG et al (1996) Effects of silicon on salinity tolerance of two barley cultivars. J Plant Nutr 19:173–183
Liang YC, Chen Q, Liu Q et al (2003) Exogenous silicon (Si) increases antioxidant enzyme activity and reduces lipid peroxidation in roots of salt-stressed barley (*Hordeum vulgare* L.). J Plant Physiol 160:1157–1164
Liang Y, Wong JW, Wei L (2005a) Silicon-mediated enhancement of cadmium tolerance in maize (*Zea mays* L.) grown in cadmium contaminated soil. Chemosphere 58:475–483
Liang YC, Zhang WH, Chen Q et al (2005b) Effects of silicon on tonoplast H^+–ATPase and H^+-PPase activity, fatty acid composition and fluidity in roots of salt-stressed barley (*Hordeum vulgare* L.). Environ Exp Bot 53:29–37
Liang YC, Hua H, Zhu Y et al (2006a) Importance of plant species and external silicon concentration to active silicon uptake and transport. New Phytol 172:63–72
Liang YC, Zhang WH, Chen Q et al (2006b) Effect of exogenous silicon (Si) on H^+-ATPase activity, phospholipids and fluidity of plasma membrane in leaves of salt-stressed barley (*Hordeum vulgare* L.). Environ Exp Bot 57:212–219
Liang Y, Sun W, Zhu Y et al (2007) Mechanisms of silicon-mediated alleviation of abiotic stresses in higher plants: a review. Environ Poll 147:422–428
Lindsay WL (1979) Chemical equilibria in soils. Wiley Interscience, NewYork
Liu Q, Zheng SJ, Lin XY (2004) Plant physiological and molecular biological mechanisms in response to aluminum toxicity. Chin J Appl Ecol 15:1641–1649
Lux A, Luxová M, Hattori T et al (2002) Silicification in sorghum (*Sorghum bicolor*) cultivars with different drought tolerance. Physiol Plant 115:87–92
Lux A, Luxova M, Abe J et al (2003a) Silicificationof bamboo (*Phyllostachys heterocycla* mitf.) root and leaf. Plant Soil 255:85–91
Lux A, Luxova M, Abe J et al (2003b) The dynamics of silicon deposition in the sorghum root endodermis. New Phytol 158:437–441
Ma JF, Takahashi E (2002) Soil, fertilizer and plant silicon research in Japan. Elsevier, Dordrecht
Ma JF, Yamaji N (2006) Silicon uptake and accumulation in higher plants. Trends Plant Sci 11:392–397
Ma JF, Goto S, Tamai K et al (2001a) Role of root hairs and lateral roots in silicon uptake by rice. Plant Physiol 127:1773–1780
Ma JF, Miyake Y, Takahashi E (2001b) Silicon as a beneficial element for crop plants. In: Snyder GH, Korndörfer GH, Datnoff LE (eds) Silicon in agriculture. Elsevier, Amsterdam, pp 17–39

Ma JF, Mitani N, Nagao S et al (2004) Characterization of the silicon uptake system and molecular mapping of the silicon transporter gene in rice. Plant Physiol 136:3284–3289

Ma JF, Nishimura K, Takahashi E (1989) Effect of silicon on the growth of rice plant at different growth stages. Soil Sci Plant Nutr 35:347-356

Ma JF, Tamai K, Yamaji N et al (2006) A silicon transporter in rice. Nature 440:688–691

Ma JF, Yamaji N, Mitani N et al (2007) An efflux transporter of silicon in rice. Nature 448:209–212

Makabe S, Kakuda K, Sasaki Y et al (2009) Relationship between mineral composition or soil texture and available silicon in alluvial paddy soils on the Shounai Plain, Japan. Soil Sci Plant Nutr 55:300–308

Matichencov VV, Bocharnikova EA (2001) The relationship between silicon and soil physical and chemical properties. In: Datnoff LE, Snyder GH, Korndörfer GH (eds) Silicon in agriculture. Elsevier, Amsterdam, pp 209–219

Matichenkov VV, Ammosova YM (1994) Effects of amorphous silica on some properties of Sody-Podzolic soils. Soil Sci 7:52–61

Matichenkov VV, Ammosova YM (1996) Effect of amorphous silica on soil properties of asod-podzolic soil. Eurasian Soil Sci 28:87–99

Matoh T, Kairusmee P, Takahashi E (1986) Salt-induced damage to rice plants and alleviation effect of silicate. Soil Sci Plant Nutr 32:295–304

Maxim LD, Niebo R, LaRosa S et al (2008) Product stewardship in wollastonite production. Inhal Toxicol 20:1199–1214

McCray JM, Mylavarapu R (2010) Sugarcane nutrient management using leaf analysis. Florida Cooperative Extension Service Fact Sheet SS-AGR-335. UF/IFAS Electronic Data Information Source (EDIS) Database. University of Florida, Gainesville. http://edis.ifas.ufl.edu/ag345. Accessed 19 July 2014

McCray JM, Rice RW, Baucum LE (2011) Calcium silicate recommendations for sugarcane on Florida organic soils. Florida Cooperative Extension Service Fact Sheet SS-AGR-350.UF/IFAS Electronic Data Information Source (EDIS) Database. University of Florida, Gainesville. http://edis.ifas.ufl.edu. Accessed 19 July 2014

McKeague JA, Cline MG (1963) Silica in soil solutions. I. The form and concentration of dissolved silica in aqueous extracts of some soils. Can J Soil Sci 43:70–82

McKeyes E, Sethi A, Young RN (1974) Amorphous coatings on particles of sensitive clay soils. Clay Miner 22:427–433

Mehra OP, Jackson ML (1960) Iron oxide removal from soils and clays by a dithionite-citrate system buffered by sodium bicarbonate. Nat Conf Clays Clay Minerals 7:317–327

Mengel K, Kirkby EA (2001) Principles of plant nutrition, 5th edn. Kluwer, Dordrecht, p 849

Menzies J, Bowen P, Ehret D et al (1992) Foliar applications of potassium silicate reduce severity of powdery mildew on cucumber, muskmelon, and zucchini squash. J Am Soc Hortic Sci 117:902–905

Meunier JD (2003) Le role des plantes dans le transfert du silicium a la surface des continents. CR Geosci 335:1199–1206

Meunier JD, Guntzer F, Kirman S et al (2008) Terrestrial plant-Si and environmental changes. Mineral Mag 72:263–267

Meyer JH, Keeping MG (2001) Past, present and future research of the role of silicon for sugar cane in southern Africa. In: Datnoff LE, Snyder GH, Korndörfer GH (eds) Silicon in agriculture. Elsevier, Amsterdam, pp 257–275

Mitani N, Ma JF (2005) Uptake system of silicon in different plant species. J Exp Bot 56:1255–1261

Monger HC, Kelly EF (2002) Silica minerals. In: Soil Mineralogy with environmental applications. Soil Science Society of America, Madison, pp 611–636

Motomura K, Fuji T, Suzuki M (2004) Silica deposition in relation to ageing of leaf tissues in *Sasa veitchii* (Carriére) Rehder (Poacceae: Bambusoideae). Ann Bot 93:235–248

Myhr K, Erstad K (1996) Converter slag a liming material on organic soils. Nor J Agric Sci 10:81

Nair PK, Aiyer RS (1968) Study of available silicon in rice soils of Keralastate (India). II. Silicon uptake by different varieties of rice in relation to available silicon contributed by soil and irrigation water. Agric Res J Kerala 6:88–99
Narayanaswamy C, Prakash NB (2009) Calibration and categorization of plant available silicon in rice soils of South India. J Plant Nutr 32:1237–1254
Neumann D, zur Nieden U (2001) Silicon and heavy metal tolerance of higher plants. Phytochemistry 56:685–692
Nonaka K, Takahashi K (1988) A method of measuring available silicates in paddy soils. Jpn Agric Res Q 22:91–95
Nonaka K, Takahashi K (1990) A method of assessing the need of silicate fertilizers in paddy soils. XIV Int Congr Soil Sci, Kyoto, Japan 4:513–514
Norton LD (1984) Micromorphology of silica cementation in soils. In: Ringrose-Voase AJ, Humphreys GS (eds) Soil micromorphology: studies in management and genesis. De Soil Sci 22:811–824
Novozamsky I, van Eck R, Houba VJG (1984) A rapid determination of silicon in plant material. Commun Soil Sci Plant Anal 15:205–211
Nowakowski W, Nowakowska J (1997) Silicon and copper interaction in the growth of spring wheat seedlings. Biol Plant 39:463–465
Nwugol CC, Huerta AJ (2008) Silicon-induced cadmium resistance in rice (*Oryza sativa*). J Plant Nutr Soil Sci 171:841–848
O'Reilly SE, Sims JT (1995) Phosphorus adsorption and desorption in a sandy soil amended with high rates of coal fly ash. Commun Soil Sci Plant Anal 26:2983
Opfergelt S, Bournonville G, Cardinal D et al (2009) Impact of soil weathering degree on silicon isotopic fractionation during adsorption onto iron oxides in basaltic ash soils, Cameroon. Geochim Cosmochim Acta 73:7226–7240
Opfergelt S, Cardinal D, André L, Delvigne C, Bremond L, Delvaux B (2010) Variations of 30Si and Ge/Si with weathering and biogenic input in tropical basaltic ash soils under monoculture. Geochim Cosmochim Acta 74:225–240
Ostatek-Boczynski ZA, Haysom MB (2003) The determination of silicon in sugarcane tissue using a microwave-assisted acid and alkali digestion procedure. In: Currie LP, Hanley JA (ed) Tools for nutrient and pollution management (Occasional report No 17). Fertilizer and Line Research Centre, Massey University, Palmerston North, pp 419–424
Osuna-Canizales FJ, DeDatta SK, Bonman JM (1991) Nitrogen form and silicon nutrition effects on resistance to blast disease of rice. Plant Soil 135:223–231
Owino-Gerroh C, Gascho GJ (2004) Effect of silicon on low pH soil phosphorus sorption and on uptake and growth of maize. Commun Soil Sci Plant Anal 35:2369–2378
Panov NP, Goncharova NA, Rodionova LP (1982) The role of amorphous silicic acid in solonetz soil processes. Vestnik Agr Sci 11:18
Parfitt RL (1978) Anion adsorption by soils and soil materials. Adv Agron 30:1–50
Park CS (2001) Past and future advances in silicon research in the Republic of Korea. In: Datnoff LE, Snyder GH, Korndörfer GH (eds) Silicon in agriculture. Elsevier Science. B.V, Amsterdam, pp 359–372
Park YS, Oh WK, Park CS (1964) A study of the silica content of the rice plant. Res Rep Rural Dev Suwon 7:31–38
Pei ZF, Ming DF, Liu D et al (2010) Silicon improves the tolerance of water-deficit stress induced by polyethylene glycol in wheat (*Triticum aestivum* L.) seedlings. J Plant Growth Reg 29:106–115
Pereira HS, Korndörfer GH, Moura WF et al (2003) Extractors of available silicon in slags and fertilizers. Rev Bras Ci Solo 27:265–274
Pervova NE, Evdokimova TI (1984) The composition of soil solutions in the subzone of Southern taiga. Pochvovedenie 1:32–39
Pokrovsky GS, Schott J, Garges F et al (2003) Iron(III)-silica interactions in aqueous solution: Insights from X-ray absorption fine structure spectroscopy. Geochim Cosmochim Acta 67:3559–3573

Ponnamperuma FN (1965) Dynamic aspects of flooded soils and the nutrition of the rice plant. In: The Mineral Nutrition of the Rice Plant. John Hopkins Press, Baltimore, pp 295–328
Prakash NB, Chandrashekar N, Mahendra C et al (2011) Effect of foliar spray of soluble silicic acid on growth and yield parameters of wetland rice in hilly and coastal zone soils of Karnataka, South India. J Plant Nutr 34:1883–1893
Rafi MM, Epstein E, Falk RH (1997) Silicon deprivation causes physical abnormalities in wheat (*Triticum aestivum* L.). J Plant Physiol 151:497–501
Raghupathy B (1993) Effect of lignite fly ash on rice. Int Rice Res Notes 18:27–28
Rahman MT, Kawamura L, Koyama H et al (1998) Varietal differences in the growth of rice plants in response to aluminum and silicon. Soil Sci Plant Nutr 44:423
Raven JA (1983) The transport and function of silicon in plants. Biol Rev 58:179–207
Reimers NF (1990) Natural uses. Dictionary-Reference Book, Moscow
Richmond KE, Sussman M (2003) Got silicon? The non-essential beneficial plant nutrient. Curr Opin Plant Biol 6:268–272
Rizwan M, Meunier J, Miche H et al (2012) Effect of silicon on reducing cadmium toxicity in durum wheat (*Triticum turgidum* L. cv. Claudio W.) grown in a soil with aged contamination. J Hazard Mater 209–210:326–334
Rodrigues FA, Duarte HSS, Domiciano GP (2009) Foliar application of potassium silicate reduces the intensity of soybean rust. Aust Plant Pathol 38:366–372
Rogalla H, Römheld V (2002) Role of leaf apoplast in silicon-mediated manganese tolerance of *Cucumis sativus* L. Plant Cell Environ 25:549–555
Saccone L, Conley DJ, Koning E et al (2007) Assessing the extraction and quantification of amorphous silica in soils of forest and grassland ecosystems. Eur J Soil Sci 58:1446–1459
Sauer D, Burghardt W (2000) Chemical processes in soils on artificial materials: silicate dissolution, occurrence of amorphous silica and zeolites. In: Proceedings of first international conference on soils of urban, industrial, traffic and mining areas, Essen,12–18 July 2000, vol 1, pp 339–346
Sauer D, Burghardt W (2006) The occurrence and distribution of various forms of silica and zeolites in soils developed from wastes of iron production. Catena 65:247–257
Sauer D, Saccone L, Conley DJ et al (2006) Review of methodologies for extracting plant-available and amorphous Si from soils and aquatic sediments. Biogeochemistry 80:89–108
Savant NK, Snyder GH, Datnoff LE (1997) Silicon management and sustainable rice production. Adv Agron 58:151–199
Savant NK, Korndörfer GH, Datnoff LE et al (1999) Silicon nutrition and sugarcane production: a review. J Plant Nutr 22:1853–1903
Schachtschabe lP, Heinemann CG (1967) Wasserlo sliche Kieselsa ure in Lo ßbo den. Z Pflanzenern Bodenk 118:22–35
Schindler PW, Furst B, Dick R et al (1976) Ligand properties of surface silanol groups. I. Surface complex formation with Fe^{3+}, Cu^{2+}, Cd^{3+}, and Pb^{2+}. J Colloid Interface Sci 55:469
Schulthess CP, Tokunaga Y (1996) Metal and pH effects on adsorption of poly(vinilalcohol) by silicon oxide. Soil Sci Soc Am J 60:92
Schwertmann U, Taylor RM (1989) Iron oxides. In: Minerals in soil environments. Soil Science Society of America, Madison, pp 379–438
Sebastian D (2012) Silica SLV results, presented at the Association of American Plant Food Control Officials Lab Services Committee Meeting, San Antonio, 21 Feb 2012. http://www.aapfco.org/meetings_aapfco.html. Accessed 8 May 2012)
Sebastian D, Rodrigues H, Kinsey C et al (2013) A 5-day method for determination of soluble silicon concentration in nonliquid fertilizer materials using a sodium carbonate-ammonium nitrate extractant followed by visible spectroscopy with heteropoly blue analysis: single laboratory validation. J AOAC Int 96:251–259
Seyfferth A, Fendorf S (2012) Silicate mineral impacts on the uptake and storage of arsenic and plant nutrients in rice (*Oryza sativa* L.). Environ Sci Tech Am Chem Soc 46:13176–13183
Shi Q, Bao Z, Zhu Z et al (2005) Silicon-mediated alleviation of Mn toxicity in *Cucumis sativus* in relation to activities of superoxide dismutase and ascorbate peroxidase. Phytochemistry 66:1551–1559

Shi GR, Cai Q, Liu C et al (2010) Silicon alleviates cadmium toxicity in peanut plants in relation to cadmium distribution and stimulation of antioxidative enzymes. Plant Growth Regul 61:45–52

Siever R, Woodford N (1973) Sorption of silica by clay minerals. Geochim Cosmochim Acta 37:1851–1880

Singh KP, Sarkar MC (1992) Phosphorus availability in soil as affected by fertilizer phosphorus, sodium silicate and farm yard manure. J Indian Soc Soil Sci 40:762–767

Sistani KR, Savant NK, Reddy KC (1997) Effect of rice hull ash silicon on rice seedling growth. J Plant Nutr 20:195–201

Snyder GH (1991) Development of a silicon soil test for Histosol-grown rice. EREC Res Rpt, EV-1991–2, University of Florida, BelleGlade

Snyder GH (2001) Methods for silicon analysis in plants, soils, and fertilizer. In: Datnoff LE, Snyder GH, Korndörfer GH (eds) Silicon in agriculture. Elsevier Science B.V, New York, pp 185–207

Snyder GH, Jones DB, Gascho GJ (1986) Silicon fertilization of rice on Everglades Histosols. Soil Sci Soc Am J 50:1259–1263

Snyder GH, Matichenkov VV, Datnoff LE (2007) Silicon. In: Barker AV, Pilbeam DJ (eds) Handbook of plant nutrition. Taylor and Francis Group/CRC Press, Boca Raton, pp 551–565

Sommer M, Kaczorek D, Kuzyakov Y et al (2006) Silicon pools and fluxes in soils and landscapes-a review. J Plant Nutr Soil Sci 169:310–329

Song Z, Wang H, Strong PJ et al (2013) Increase of available silicon by Si-rich manure for sustainable rice production. Agron Sustain Dev. doi:10.1007/s13593-013-0202-5

Sparkman OD (2006) Mass spectrometry desk reference. Global View Publishing, Pittsburgh

Street-Perrott FA, Barker P (2008) Biogenic silica: a neglected component of the coupled global continental biogeochemical cycles of carbon and silicon. Earth Surf Proc Land 33:1436–1457

Sun L, Gong K (2001) Silicon-based materials from rice husks and their applications. Ind Eng Chem Res 40:5861–5877

Taber HG, Shogren D, Lu G (2002) Extraction of silicon from plant tissue with dilute HCl and HF and measurement by modified inductive coupled argon plasma procedures. Commun Soil Sci Plant Anal 33:1661–1670

Takahashi K (1981) Effects of slags on the growth and the silicon uptake by rice plants and the available silicates in paddy soils. Bull Shikoku Natl Agric Exp Stn 38:75–114

Takahasi E, Ma JF, Miyake Y (1990) The possibility of silicon as an essential element for higher plants. Comments Agric Food Chem 2:99–122

Takijima Y, Wijayaratna HMS, Seneviratne CJ (1970) Nutrient deficiency and physiological disease of lowland rice in Ceylon. III. Effect of silicate fertilizers and dolomite for increasing rice yields. Soil Sci Plant Nutr 16:11–16

Tan KH (1994) Environmental soil science. Marcel Dekker, New York, pp 193–198

Tokunaga Y (1991) Potassium silicate: a slow release potassium fertilizer. Fert Res 30:55

Tubaña B, Narayanaswamy C, Datnoff LE et al (2012a) Changes in pH and extractable nutrients of selected soils from the Midwest and South USA as influenced by different rates of iron calcium silicate slag. ASA-CSSA-SSSA International Annual Meetings, Cincinnati, 21–24 Oct

Tubaña B, Narayanaswamy C, Kanke Y et al (2012b) Estimation of plant available silicon using different extraction procedures for selected soils from the Midwest and South USA.ASA-CSSA-SSSA International Annual Meetings, Cincinnati, 21–24 Oct

Van Cappellen P (2003) Biomineralization and global biogeochemical cycles. Rev Miner Geochem 54:357–381

Van der Vorm PDJ (1987) Dry ashing of plant material and dissolution of the ash in HF for the colorimetric determination of silicon. Commun Soil Sci Plant Anal 18:1181–1189

Varela-Milla O, Rivera EB, Huang WJ et al (2013) Agronomic properties and characterization of rice husk and wood biochars and their effect on the growth of water spinach in a field test. J Soil Sci Plant Nutr 13:251–266

Vasanthi ND, Saleena LM, Raj SA (2012) Silicon in day today life. World Appl Sci J 17:1425–1440

Volker H, Nudling W, Adam K (1985) Verfahren zur herstellung eines mehrnahrstofffdungers. Fels-werke Peine-Salzgitter GmBH, Pat. 3538411.5 Germany
Volkova VV (1980) Silicate content in soil solutions and natural waters of the Russian plain. In: Kovda VA (ed) Pedological and biogeocenotic research of the Russian lowland centre. Nauka Publication, Moscow
Wada K (1989) Allophane and imogolite. In: Minerals in soil environments. Soil Science Society of America, Madison, pp 1051–1087
Wada K, Greenland DJ (1970) Selective dissolution and differential infrared spectroscopy for characterization of 'amorphous' constituents in soil clays. Clay Miner 8:241–254
Wallace A (1993) Participation of silicon in cation-anion balance as possible mechanism for aluminum and iron tolerance in some *gramineae*. J Plant Nutr 16:547–553
Wang L, Wang W, Chen Q et al (2000) Silicon induced cadmium tolerance of rice seedlings. J Plant Nutr 23:1397–1406
Wang Y, Stass A, Horst WJ (2004) Apoplastic binding of aluminum is involved in silicon-induced amelioration of aluminum toxicity in maize. Plant Physiol 136:3762–3770
Weast RC, Aastle MJ, Beyer WH (1985) Handbook of chemistry and physics, 65th edn. CRC Press, Boca Raton
Wedepohl KH (1995) The composition of the continental crust. Geochim Cosmochim Acta 59:1217–1232
Williams LA, Crerar DA (1985) Silica diagenesis. II. General mechanisms. J Sed Pet 55:312–321
Winslow MD (1995) Silicon: A new macronutrient deficiency in upland rice. Working document No 10. International Center of Tropical Agriculture, Cali
Wonisch H, Gerard F, Dietzel M et al (2008) Occurrence of polymerized silicic acid and aluminum species in two forest soil solutions with different acidity. Geoderma 144:435–445
Xu G, Zhan X, Li C et al (2001) Assessing methods of available silicon in calcareous soils. Commun Soil Sci Plant Anal 32:787–801
Yamaji N, Mitani N, Ma JF (2008) A transporter regulating silicon distribution in rice shoots. Plant Cell 20:1381–1389
Yates DM, Joyce KJ, Heaney PJ (1998) Complexation of copper with polymeric silica in aqueous solution. Appl Geochem 13:235–241
Yeo AR, Flowers SA, Rao G et al (1999) Silicon reduces sodium uptake in rice (*Oryza sativa* L.) in saline conditions and this is accounted for by a reduction in the transpirational bypass flow. Plant Cell Environ 22:559–565
Yoshida S, Ohnishi Y, Kitagishi K (1962) Chemical forms, mobility, and deposition of silicon in the rice plant. Soil Sci Plant Nutr 8:107–111
Yoshida S, Forno DA, Cock JH et al (1976) Laboratory manual for physiological studies of rice, 3rd edn. International Rice Research Institute, Los Banos, pp 17–22
Zhu ZJ, Wei GQ, Li J et al (2004) Silicon alleviates salt stress and increases antioxidant enzymes activity in leaves of salt-stressed cucumber (*Cucumis sativus* L.). Plant Sci 167:527–533

Chapter 3
Silicon Control of Soil-borne and Seed-borne Diseases

Alessandro Antônio Fortunato, Fabrício A. Rodrigues, and Lawrence E. Datnoff

Abstract The severity of several plant diseases caused by both soil-borne and seed-borne pathogens are dramatically decreased in agronomic and horticultural crops when they are produced in soils or in hydroponic culture amended with soluble silicon. Wilts, root rots, and galling caused by plant pathogens, such as *Fusarium, Pythium, Rhizoctonia*, and *Meloidogyne*, are less severe when silicon is made available resulting in slower disease progress and less disease severity. Brown spot, caused by *Bipolaris oryzae*, a devastating seed-borne disease of rice, causes severe grain discoloration that can be suppressed using silicon. A significant decrease in grain and seedling discoloration was observed for plants supplied with silicon. In addition, seedling emergence was vastly improved. Because host resistance to these diseases may be limited, and the efficacy of fungicide applications may be erratic, at best, for suppressing both soil- and seed-borne diseases, silicon undoubtedly is a well-suited strategy for inclusion in an integrated disease management program.

Introduction

The occurrence of soil-borne diseases, such as those causing damping-off and crown and root rot, in high-value horticultural crops is one of the major obstacles that greatly contribute to a decrease in crop quality and yield. Pathogens that infect and/or infest seeds of important agricultural crops also represent a major threat to crop establishment. Sowing seeds free of plant pathogens is a key management strategy to prevent the introduction of plant diseases, especially new ones, into a

A.A. Fortunato (✉) • F.A. Rodrigues
Departamento de Fitopatologia, Universidade Federal de Viçosa, Viçosa, Minas Gerais Cep 36570-900, Brazil
e-mail: fabricio.plantpathologist@gmail.com; fabricio@ufv.br

L.E. Datnoff
Department of Plant Pathology and Crop Physiology, Louisiana State University Agricultural Center, 302 Life Sciences, Baton Rouge, LA 70803, USA
e-mail: LDatnoff@agcenter.lsu.edu

F.A. Rodrigues, L.E. Datnoff (eds.), *Silicon and Plant Diseases*,
DOI 10.1007/978-3-319-22930-0_3

production field. Indeed, the widespread distribution of diseases within the crop is maximized when infected and/or infested seeds are used, which will then contribute to a high number of initial infection sites and to subsequently higher plant disease epidemic rates. The management of soil- and seed-borne diseases is particularly challenging because many causal agents will survive in the soil for several years through the production of resistant structures, such as chlamydospores and/or sclerotia. Some soil-borne and seed-borne pathogens that infect either monocots or dicots exhibit a decrease in their disease intensities when plants have an adequate tissue level of silicon (Datnoff et al. 2007). The effect of silicon on the development of these plant diseases will be discussed in greater depth in this chapter. A detailed list of plant diseases caused by both soil-borne and seed-borne pathogens in agronomic and horticultural crops that have had their intensities reduced by silicon is summarized in Table 3.1.

Table 3.1 Effect of silicon on some soil-borne and seed-borne diseases

Hosts	Diseases	Pathogens	Effects[a]	References
Avocado	Phytophthora root rot	*Phytophthora cinnamomi*	⊕	Bekker et al. (2005)
Banana	Root rot	*Cylindrocladium spathiphylli*	⊕	Vermeire et al. (2011)
	Panama disease	*Fusarium oxysporum* f. sp. *cubense*	⊕	Fortunato et al. (2012)
	Root-knot nematode	*Meloidogyne javanica*	⊕	Oliveira et al. (2012)
Bell pepper	Phytophthora blight	*Phytophthora capsici*	⊕	Lee et al. (2004), French-Monar et al. (2010)
Bitter gourd	Pythium root rot	*Pythium aphanidermatum*	⊕	Heine et al. (2007)
Coffee	Root-knot nematode	*Meloidogyne exigua*	⊕	Silva et al. (2010)
Corn	Pythium root rot	*Pythium aphanidermatum*	⊕	Sun et al. (1994)
	Stalk rot	*Fusarium moniliforme*	⊕	
Creeping bentgrass	Pythium root rot	*Pythium aphanidermatum*	⊕	North Carolina State University (1997), Schmidt et al. (1999), Rondeau (2001), Uriarte et al. (2004), Zhang et al. (2006)
	Dollar spot	*Sclerotinia homoeocarpa*	⊕	
	Brown patch	*Rhizoctonia solani*	⊕	
Cucumber	Crown and root rot	*Pythium ultimum*	⊕	Chérif and Bélanger (1992)
	Crown and root rot	*Pythium aphanidermatum*	⊕	Chérif et al. (1994)
	Fusarium wilt	*Fusarium oxysporum* f. sp. *cucumerinum*	⊕	Miyaki and Takahashi (1983)

(continued)

Table 3.1 (continued)

Hosts	Diseases	Pathogens	Effects[a]	References
Lettuce	Fusarium wilt	*Fusarium oxysporum* f. sp. *lactucae*	⊕	Chitarra et al. (2013)
Melon	Fusarium root rot	*Fusarium* spp.	⊕	Liu et al. (2009)
Oil palm	Basal stem rot	*Ganoderma boninense*	⊕	Najihah et al. (2015)
Perennial ryegrass	Fusarium patch	*Microdochim nivale*	⊕	MacDonagh and Hunter (2010)
Rice	Root knot nematodes	*Meloidogyne* spp.	⊕	Swain and Prasad (1988)
	Grain discoloration	Many fungal species	⊕	Winslow (1992), Korndörfer et al. (1999), Prabhu et al. (2012), Dallagnol et al. (2013, 2014)
Soybean	Phytophthora root rot	*Phytophthora sojae*	⊕	Guérin et al. (2014)
Tall fescue	Brown patch	*Rhizoctonia solani*	⊗	Zhang et al. (2006)
Tomato	Fusarium wilt	*Fusarium oxysporum* f. sp. *lycopersici* races 1 and 2	∅	Rodrigues et al. (1996)
	Fusarium crown and root rot	*Fusarium oxysporum* f. sp. *radices-lycopersici*	∅	Menzies et al. (2001)
	Fusarium crown and root rot	*Fusarium oxysporum* f. sp. *radices-lycopersici*	⊕	Huang et al. (2011)
	Pythium root rot	*Pythium aphanidermatum*	⊕	Heine et al. (2007)
	Bacterial wilt	*Ralstonia solanacearum*	⊕	Dannon and Wydra (2004), Kiirika et al. (2013)
Watermelon	Gummy stem blight	*Didymella bryoniae*	⊕	Santos et al. (2010)
Wheat	Foot rot	*Fusarium* spp.	⊕	Rodgers-Gray and Shaw (2000; 2004)
Zoysiagrass	Brown patch	*Rhizoctonia solani*	⊕	Saigusa et al. (2000)

[a]Silicon can decrease (⊕), increase (⊗) or has no effect (∅) on disease intensity

Diseases Caused by Soil-borne Pathogens

Lee and his colleagues (2004) demonstrated that the level of resistance (root decay and the number of wilted plants) to Phytophthora blight (*Phytophthora capsici*) increased when bell pepper plants were grown in a nutrient solution containing 100 and 200 ppm of silicon, using potassium silicate as its source. French-Monar et al. (2010) also investigated the potential of silicon to decrease the symptoms of Phytophthora blight on bell pepper plants. The authors observed a 40 % increase in the silicon concentration in the roots, but not in the stems of plants supplied with silicon compared to those not supplied. The area under diseased plant progress curve and the area under wilting plant progress curve were reduced from 15 % to 38 % and from 29 % to 33 %, respectively, for plants supplied with silicon compared to those not supplied. In fact, the relative lesion extension, obtained as the ratio of vertical lesion extension to stem length, was reduced by 35 %, and the dry root and stem weights increased by 24 % and 102 %, respectively, for plants supplied with silicon compared to those not supplied. Guérin et al. (2014) investigated the potential of silicon to increase the resistance of different soybean cultivars and transgenic lines expressing silicon transporters (*TaLsi1* and *EaLsi1* genes from wheat and horsetail, respectively) to Phytophthora stem and root rot caused by *Phytophthora sojae*. According to these authors, plants from all genotypes were able to uptake and accumulate silicon in the shoots compared to those not supplied. Plants from the cultivar Hikmoksorip had significantly more silicon (twofold increase) in their shoots compared to all the other genotypes. Moreover, the transgenic lines containing the *TaLsi1* and *EaLsi1* genes did not accumulate significant amounts of silicon in the shoot compared to the soybean cultivars Hikmoksorip and Jack. Although the incubation period of Phytophthora stem and root rot was not affected by silicon, plants not supplied with silicon wilted earlier. They also had the greatest disease intensities and developed extensive symptoms of stem canker compared to those plants supplied with silicon. Plants from the transgenic lines and from the cultivar Hikmoksorip supplied with silicon had a significantly lower area under disease progress curve than non-amended plants. In the absence of silicon, plants from the cultivar Hikmoksorip were most susceptible to Phytophthora stem and root rot compared to plants from the transgenic lines containing the *TaLsi1* and *EaLsi1* genes. Plants from the cultivar Jack showed the highest level of resistance to *P. sojae* regardless of being amended or not with silicon. The area under disease progress curve for plants from the transgenic lines containing the *TaLsi1* and *EaLsi1* genes supplied with silicon was reduced to the same level as that observed for plants from the cultivar Jack. In a different study, Bekker et al. (2005) observed that avocado seedlings soaked in a solution containing 20.7 % silicon dioxide for 10 days before being inoculated by *P. cinnamomi* showed reductions in disease severity and the highest root mass compared to plants that received the solution 4 days after pathogen inoculation.

Chérif and Bélanger (1992) reported that cucumber plants supplied with 1.7 mM potassium silicate (100 ppm of silicon) showed a significant decrease in Pythium root rot (*Pythium ultimum*) decay and mortality compared to those plants not supplied. There was a significant reduction in the incidence of Pythium crown and root rot (*Pythium aphanidermatum*) for cucumber plants grown in a nutrient solution supplied with silicon compared to those plants not supplied (Chérif et al. 1994). The incidence of stalk rot caused by *P. aphanidermatum* and *Fusarium moniliforme* for corn plants supplied with silicon was greatly reduced compared to those plants not supplied (Sun et al. 1994). Silicon was effective in reducing the intensities of root rot (*P. aphanidermatum*), dollar spot (*Sclerotinia homoeocarpa*), and brown patch (*Rhizoctonia solani*) on creeping bentgrass (North Carolina State 1997; Schmidt et al. 1999; Rondeau 2001; Uriarte et al. 2004) as well as brown patch (*R. solani*) on zoysia grass (Saigusa et al. 2000). Zhang et al. (2006) verified a negative effect of silicon, using calcium silicate as its source, as a topdressing (2440 or 4880 kg/ha) for controlling brown patch (*R. solani*) on tall fescue and on creeping bentgrass in the field or when soil was amended at product rates of 7325 and 14,650 kg/ha to control dollar spot on creeping bentgrass under controlled conditions. The area under disease progress curve for brown patch increased from 26 % to 30 % for plants grown in soil amended with calcium silicate at a product rate of 2440 kg/ha. A calcium silicate topdressing increased both root and foliar silicon concentrations for creeping bentgrass plants. A positive correlation was observed between the high foliar silicon concentration and brown patch severity for creeping bentgrass plants. Despite the higher native soil silicon concentration originally observed (173 mg Si/kg), amending a silty clay loam soil (pH 7.0) with calcium silicate before planting creeping bentgrass plants did not contribute to an increase in the foliar silicon concentration nor did it reduce dollar spot incidence or brown patch severity. According to these authors, the soil amended with calcium silicate, which would release an adequate level of soluble silicon, was not very effective in suppressing brown patch development in tall fescue plants. In fact, under high natural silicon soil concentrations, calcium silicate was not a viable option for reducing the severity of brown patch or dollar spot on creeping bentgrass.

Heine et al. (2007) observed that the growth of *P. aphanidermatum* in the root apical meristem of bitter gourd plants supplied with silicon was not reduced. However, the continuous supply of silicon, especially before inoculation by *P. aphanidermatum*, significantly decreased the subapical and basipetal spread of the pathogen from the infected root apex. Therefore, based on this host-pathogen interaction, silicon must be applied continuously before root or plant inoculation or the effect of silicon in suppressing this disease will be ineffective. According to the authors, symplastic silicon played a pivotal role in decreasing *P. aphanidermatum* root colonization into the subapical and basipetal root regions.

Banana plants grown on Vertisols, which are rich in soil-soluble silicon, are in general less affected by plant pathogens compared to when plants are grown in ferrallitic soils, which are considered silicon-deficient (Henriet et al. 2006). Vermeire et al. (2011) reported a reduction of approximately 50 % in banana root necrosis due to infection by *Cylindrocladium spathiphylli* when supplied with 2 mM monosilicic acid compared to those plants not supplied. The reduction in plant growth due to infection by the pathogen was lessened and the amount of necrotic roots was lower for silicon-amended plants.

The number of cucumber plants that wilted and died due to an infection by *Fusarium oxysporum* f. sp. *cucumerinum* decreased when supplied with silicon (Miyaki and Takahashi 1983). The use of silicon oxide and sodium silicate applied as a dipping solution for melon fruits significantly reduced the disease severity caused by *Fusarium* spp. (Liu et al. 2009). However, during the *in vitro* tests, sodium silicate at 100 mM was more efficient in reducing the radial fungal mycelial growth of *Fusarium* compared to silicon oxide, which was demonstrated to be ineffective. Based on scanning electron microscopy coupled with energy-dispersive X-ray analysis, the authors observed that a high intensity of silicon deposition occurred in the epidermis, especially at the stomata, as well as at the junction between the exocarp and mesocarp, of the melon fruits (Liu et al. 2009). Rodgers-Gray and Shaw (2000; 2004) demonstrated that the severity of foot rot (*Fusarium* spp.) on wheat plants supplied with silicon was reduced by 24 % compared to the non-supplied plants. Fortunato et al. (2012) determined the effect of silicon in reducing the symptoms of Fusarium wilt (*Fusarium oxysporum* f. sp. *cubense* (*Foc*)) development in banana seedlings from the cultivars Grand Nain (resistant) and Maçã (susceptible). The silicon concentration in the roots and in the rhizome-pseudostem significantly increased by 30 % and 59 %, respectively, for silicon-supplied plants compared to those not supplied. The silicon concentration in the roots and rhizome-pseudostem of Grand Nain seedlings was 12 % and 37 % greater, respectively, than that in the seedlings from the cultivar Maçã. At 32 days after inoculation, the visual effect of supplying silicon to seedlings in suppressing Fusarium wilt development in both cultivars was noticeable. The silicon-supplied seedlings showed a significant reduction of 12 %, 50 %, 52 % and 21 %, respectively, for the area under reflex leaf symptom progress curve, the area under root symptom progress curve, the area under disease progress curve and the area under asymptomatic fungal colonization of tissue progress curve compared to those of plants not supplied. The area under darkening of rhizome-pseudostem progress curve (AUDRPPC) of Maçã seedlings increased significantly by 16 % for seedlings not supplied with silicon compared to those amended with silicon. Silicon enhanced the resistance for the susceptible cultivar, which had a significant decrease in AUDRPPC by 21 %, compared to the seedlings from the resistant cultivar. In fact, the colonization of *Foc* inside the tissues was limited by silicon, as assessed by the relative lesion length, and these silicon-amended seedlings showed a substantial reduction in the area under relative lesion length progress curve by 42 % compared to those not supplied. Due to its natural genetic resistance to *Foc*, no difference in the response to fungal infection was detected between the cultivar Grand Nain seedlings amended with or without silicon regardless of the variables used to assess disease development. Huang et al. (2011) investigated the effectiveness of silicon via sand culture fertilized with a nutrient solution amended (100 mg/L sodium metasilicate nonahydrate) or not amended with silicon to reduce the severity of Fusarium crown and root rot (*Fusarium oxysporum* f. sp. *radices-lycopersici*) of tomato plants. Four weeks after fungal inoculation, the silicon concentration in the roots and shoots of tomato plants supplied with silicon was significantly higher than that in those plants not supplied. Moreover, the increase in the root silicon concentration was significantly correlated with a reduction in disease severity on the roots, crown, and stems. Based on disease progress, the decrease in Fusarium crown and root rot severity on plants supplied with silicon was due to a delay in the onset of fungal infection in the roots and the subsequent upward movement of the fungus in the direction of the stem (Huang et al. 2011).

Chitarra et al. (2013) tested the effect of silicon on controlling Fusarium wilt (*Fusarium oxysporum* f. sp. *lactucae*) in lettuce plants grown in a soilless system not amended or amended with potassium silicate (100 mg/L) at three levels of electrical conductivity (1.5–1.6, 3.0–3.2 and 4–4.2 mS/cm). The Fusarium wilt severity was significantly reduced by 62 % in plants supplied with silicon compared to those not supplied only at the electrical conductivity of 4–4.2 mS/cm. The control of Fusarium patch (*Microdochium nivale*) on perennial ryegrass plants using silicon, applied as potassium metasilicate at 1200 ppm, via drench or foliarly was investigated by McDonagh and Hunter (2010). According to these authors, silicon applied by drenching was more effective in increasing the silicon shoot concentration compared to that applied foliarly. However, the drenching decreased the Fusarium patch incidence by approximately 18 %, while foliar application decreased the incidence by 40 %. Furthermore, foliar application prevented the coalescence of the infected patches.

Najihah et al. (2015) evaluated the potential of five silicon sources (silicon oxide, potassium silicate, calcium silicate, sodium silicate and sodium metasilicate) at four concentrations (0, 800, 1200 and 2000 mg/L) to control basal stem rot (*Ganoderma boninense*) on oil palm seedlings. For seedlings supplied with silicon oxide, infection by *G. boninense* was greatly reduced compared to non-supplied seedlings. Seedlings supplied with all silicon sources showed no external symptoms of basal stem rot for the first 2 months of fungal inoculation in contrast to the seedlings not supplied with silicon. Indeed, the emission of primary roots for seedlings supplied with silicon was greatly reduced and, consequently, so were the number of infected roots and stem lesions. Eight months after fungal inoculation, seedlings not supplied with silicon showed disease severities greater than 90 % (Najihah et al. 2015).

Santos et al. (2010) studied the influence of different silicon sources and rates (soil amendment with calcium and magnesium thermophosphate at rates of 250, 500, 1000, 2000 and 3000 kg/ha; soil amendment with calcium and magnesium silicate applied in the furrow at rates of 25, 50, 100, 200 and 300 kg/ha; and foliar sprays of potassium silicate at rates of 250, 500, 1000, 1500 and 2500 ml/ha) and the control treatment (plants grown in non-amended soil with silicon-source products or plants not sprayed with potassium silicate) for controlling gummy stem blight (*Didymella bryoniae*) on watermelon. Plants amended with thermophosphate (3000 kg/ha) had the severity of gummy stem blight reduced significantly by 35 % compared to those not amended. The severity of gummy stem blight was also reduced by 12 % and 31 % for plants sprayed with potassium silicate (2500 ml/ha) and for plants grown in soil amended with calcium and magnesium silicate (300 kg/ha), respectively.

Dannon and Wydra (2004) studied the effect of silicon on enhancing the resistance of different tomato genotypes (L390, King Kong 2 and Hawaii 7998, which are susceptible, moderately resistant, and resistant, respectively) to bacterial wilt (*Ralstonia solanacearum*). These authors observed a delay in the onset of wilt symptoms caused by *R. solanacearum* in all the genotypes supplied with silicon. Moreover, the development of disease severity and wilt incidence, both expressed as area under disease progress curve, of silicon-treated plants were significantly lower compared to those not treated for the genotype L390 at 16.1 % and 26.8 %, respectively. For the genotype King Kong 2, disease development was slower for plants supplied with silicon than for non-supplied plants, and wilt incidence development was delayed by 6 days. L390 and King Kong 2 supplied with silicon showed a reduction of 27 % and 56 %, respectively, for area under disease progress curve compared to plants not supplied with silicon. Plant death did not occur until 12 days after inoculation for those plants supplied with silicon, while at 11 days after inoculation, 63 % of the plants not supplied with silicon already had died. At the end of the experiments, 46 % of the plants supplied with silicon survived, in contrast to 33 % of those not supplied. No symptoms of wilt were observed for the genotype Hawaii 7998 whether amended or not with silicon. Kiirika et al. (2013) investigated the combination of chitosan and silicon for controlling bacterial wilt (*R. solanacearum*) in tomato plants for the genotypes King Kong and L390. According to these authors, bacterial wilt incidence decreased by 40 % and 57 % for King Kong 2 and by 27 % and 33 % for L390 treated with silicon and chitosan, respectively. The combination of silicon and chitosan significantly reduced bacterial wilt incidence by 75 % and 47 % for King Kong 2 and L390, respectively.

Rice cultivars with a high silicon concentration in the roots were more resistant to the root-knot nematode *Meloidogyne* spp. (Swain and Prasad 1988). The coffee cultivars Catuaí 44 and IAPAR 59, susceptible and resistant, respectively, to the root-knot nematode *Meloidogyne exigua*, were grown in pots containing a silicon-deficient soil amended with either calcium silicate or calcium carbonate (Silva et al. 2010). There was an increase of 152 % and 100 %, respectively, in the root silicon concentration for Catuaí 44 and IAPAR 59 plants grown in the presence of calcium silicate compared to those grown in the presence of calcium carbonate. In addition, no significant differences were detected in the root calcium concentration between calcium silicate and calcium carbonate treatments. The number of galls and the number of eggs of *M. exigua* significantly decreased by 17 % and 28 %, respectively, for plants from Catuaí 44 amended with calcium silicate. The number of galls and the number of eggs were significantly lower for IAPAR 59 plants compared to those for Catuaí 44 regardless of whether the plants were grown in soil amended with calcium silicate or calcium carbonate.

Diseases Caused by Seed-borne Pathogens

The application of 18.7 g of silicon/m^2, using sodium metasilicate as its source, on highly weathered Ultisols in West Africa cultivated with upland rice doubled the foliar silicon concentration and significantly reduced grain husk discoloration (Winslow 1992). Korndörfer et al. (1999) evaluated the effects of silicon on grain discoloration in rice grown in four different soil types low in plant-available silicon. These authors verified that silicon amendments to rice plants reduced grain discoloration regardless of the soil type. Seebold and his colleagues (2000) demonstrated that silicon (1000 kg/ha) reduced grain discoloration by 25.6 % and 68.4 % at two field locations in eastern Colombia. These authors further observed that head rice (whole grains) increased by 20.5 % and 25.7 %. Prabhu et al. (2012) investigated the potential of different silicon rates on the reduction of grain discoloration (*Bipolaris oryzae*) in several rice genotypes. These authors observed that a negative quadratic relationship existed between silicon rates and grain discoloration for 48 rice genotypes. Although the silicon concentration increased in the husk for all genotypes, no relationship was detected between the grain discoloration and silicon concentration regardless of the silicon rates. Furthermore, genotypic differences in silicon concentration in the husk were noted regardless of the silicon rates. Dallagnol et al. (2013) investigated the importance of silicon in rice grain husk along with fungicide applications to prevent the transmission of *B. oryzae* from seeds to seedlings and to improve seedling emergence. Plants from the cultivar Oochikara and the mutant *lsi1* (defective in the *Lsi1* transporter for silicon uptake) were grown in a nutrient solution without or with silicon, and at the milk-grain growth stage, their panicles were inoculated with *B. oryzae*. Seeds were evaluated for brown spot severity and husk silicon concentration. The silicon concentration in the husks of plants from the cultivar Oochikara was up to four times higher than that of the *lsi1* mutant. The severity of brown spot on the grain was reduced when the plants were supplied with silicon (Fig. 3.1). For the cultivar Oochikara supplied with silicon, a high percentage (79.8 %) of symptomatic grains had severities less than 10 % and received a score of 2 (from 1.1 % to 5 % of the grain area showing disease symptoms), followed by scores of 1 and 3 (less than 1 % and from 5.1 % to 10 %, respectively, of the grain area showing disease symptoms) (Fig. 3.1); only 1.3 % of the grains were rated with a score of 6 (from 50.1 % to 75 % of the grain area showing disease symptoms). In contrast, the *lsi1* mutant plants supplied with silicon had a high percentage (75.7 %) of symptomatic grains with scores of 1–4 (from 10.1 % to 25 % of the grain area showing disease symptoms) and had an increase in the percentage of grains with severities greater than 10 % (Fig. 3.1). Grains from plants supplied with silicon had significantly greater percentages of scores of 1 and 2, and significantly lower percentages of scores of 5 (from 25.1 % to 50 % of the grain area showing disease symptoms), 6 and 7 (more than 75 % of the grain area showing disease

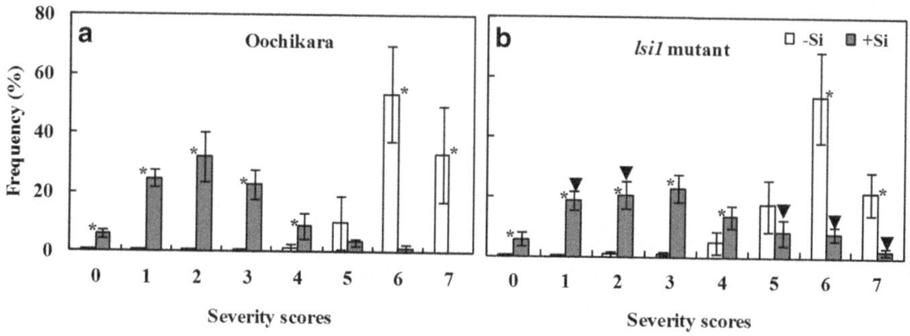

Fig. 3.1 Frequency of brown spot severity scores for rice seeds from the cultivar Oochikara (**a**) and *lsi1* mutant (**b**) plants grown in a hydroponic culture containing 0 (−Si) or 2 (+Si) mM silicon (Si) and inoculated with *Bipolaris oryzae*. Means for each plant type from the −Si and +Si treatments followed by an asterisk (*) are significantly different from each other based on the *t*-test ($P \leq 0.05$). Means from the *lsi1* mutant for −Si and +Si treatments followed by an inverted triangle (▼) are significantly different from the cultivar Oochikara based on the *t*-test ($P \leq 0.05$). ($n = 10$) (This figure is reproduced with permission from Tropical Plant Pathology 38:478–484, 2013)

symptoms) was found for plants from the cultivar Oochikara compared to those from the *lsi1* mutant (Fig. 3.1). In contrast, no significant difference in brown spot severity on the grains was observed between the cultivar Oochikara and the *lsi1* mutant in the absence of silicon. High percentages of symptomatic grains from the cultivar Oochikara and the *lsi1* mutant without silicon (97 % and 93 %, respectively) were rated with scores of 5–7; none of the grains were rated with a score of 0 (no disease symptoms), and only a few grains received scores of 1–3 (0.9 % and 2 % for cultivar Oochikara and the *lsi1* mutant, respectively). For the cultivar Oochikara supplied with silicon, the grains were ranked into classes 1 (grains that received scores of 0–5) and 2 (grains that received scores of 6 and 7) at 98.7 % and 1.3 %, respectively, whereas 89 % and 11 % of the grains obtained from the *lsi1* mutant supplied with silicon were ranked into classes 1 and 2, respectively. For the cultivar Oochikara and the *lsi1* mutant not supplied with silicon, 12.7 % and 23.6 % of the grains were ranked into class 1, respectively, and 87.3 % and 76.4 % were ranked into class 2, respectively. A higher percentage of seedling emergence occurred from seeds obtained from plants supplied with silicon, particularly for the cultivar Oochikara, and a lower percentage of infected seedlings occurred from seeds obtained from the cultivar Oochikara supplied with silicon, particularly when these seeds were treated with fungicide. The fungicide was not very effective in preventing seedlings from being infected by *B. oryzae* from seeds with brown spot severity ratings greater than 50 %. A silicon concentration in the husk greater than 3 dag/kg resulted in a lower brown spot severity rating, and the fungicide efficacy was greater for a low brown spot severity rating on seeds from plants supplied with silicon, particularly for the cultivar Oochikara. A low fungicide efficacy was observed on seeds from plants not supplied with silicon,which received a higher seed brown spot severity rating. As a consequence, the fungicide treatment was not very effective in preventing fungal infection in the seedlings. The importance of silicon in rice husks showed that seed health and physiology improved based on the following: a higher percentage of emerged seedlings, a lower percentage of infected seedlings and a higher seedling dry matter weight, as observed for the cultivar Oochikara supplied with silicon (Dallagnol et al. 2013). In another study, Dallagnol et al. (2014) grew rice plants from the cultivar Oochikara and the *lsi1* mutant in a hydroponic culture with and without silicon. At the beginning of the milk-grain growth stage, the panicles of Oochikara and the *lsi1* mutant were inoculated with *B. oryzae* and were harvested afterward at physiological grain maturity to determine the effect of silicon on grain resistance to brown spot development. The supply of silicon significantly increased the silicon concentration in husks compared to those not supplied. The silicon concentration in husks from the cultivar Oochikara was up to three times greater than that in the *lsi1* mutant. In the presence of silicon, brown spot severity was reduced by 88 % in grains from the cultivar Oochikara and by 53 % in grains from the *lsi1* mutant. Brown spot severity was 77 % lower for grains of the cultivar Oochikara than for the *lsi1* mutant when both plant types were grown in the presence of silicon. Brown spot severity was reduced in the husks of the seeds

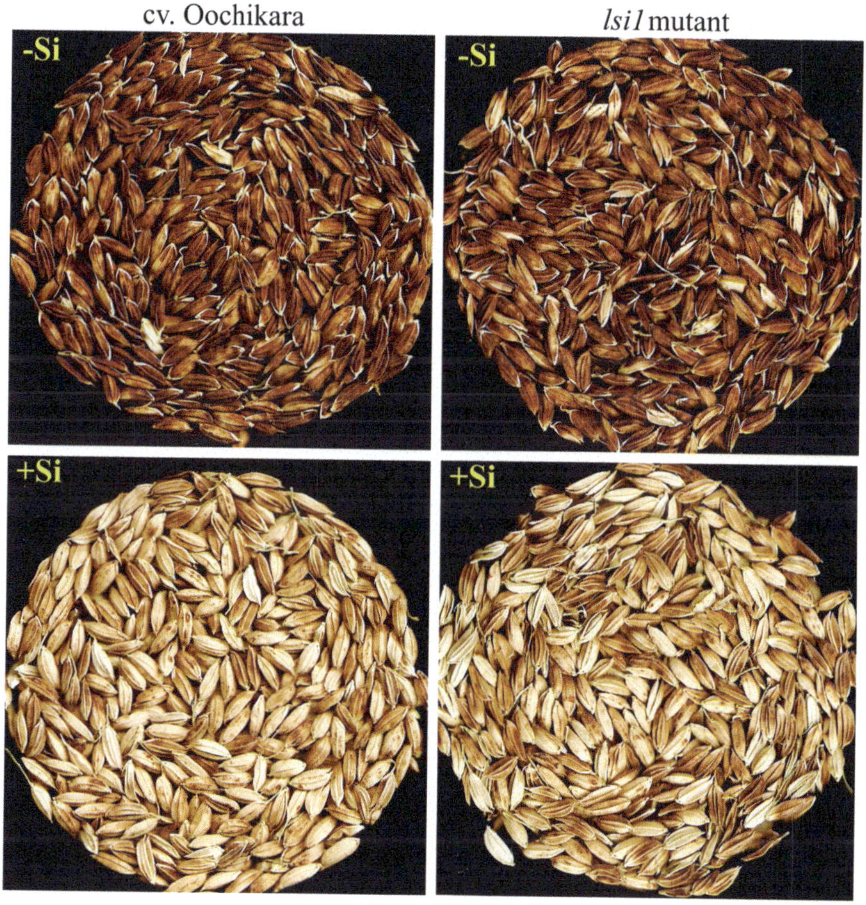

Fig. 3.2 Symptoms of brown spot on the grains of the cultivar Oochikara and *lsi1* mutant plants grown in hydroponic culture either without (–Si) or with (+Si; 2 mM) silicon (Si) and inoculated with *Bipolaris oryzae*

from plants supplied with silicon, particularly for the cultivar Oochikara (Fig. 3.2). Panicle inoculation significantly reduced the following yield components: the number of grains per panicle, the 1000-grain weight and the percentage of filled grains. In the presence of silicon, these yield components significantly increased, especially for inoculated panicles. Considering the kernel quality, panicle inoculation with *B. oryzae* significantly reduced the yield of the husked kernel, the yield of the whole kernel, and the kernel diameter, especially for grains from plants not supplied with silicon. For panicles from plants supplied with silicon, the kernel quality was greatly improved compared to those not supplied, especially when inoculated with *B. oryzae* (Dallagnol et al. 2014).

Conclusions

The negative impact caused by soil- and seed-borne diseases on the quality and yield of crops of economic importance may be reduced by simply adding silicon to the soil or nutrient solution. Because host resistance may be limited, and fungicide efficacy may be erratic at best for suppressing both soil- and seed-borne diseases, silicon is clearly a well-suited strategy for inclusion in an integrated disease management program.

References

Bekker T, Labuschagne N, Kaiser C (2005) Effects of soluble silicon against *Phytophthora cinnamomi* root rot of avocado (*Perseaamericana* Mill). nursey plants. S Afr Avocado Grow Assoc Yearb 28:60–64

Chérif M, Bélanger RR (1992) Use of potassium silicate amendments in recirculating nutrient solutions to suppress *Pythium ultimum* on long english cucumber. Plant Dis 76:1008–1011

Chérif M, Asselim A, Bélanger RR (1994) Defense responses induced by soluble silicon in cucumber roots infected by *Pythium* spp. Phytopathology 84:236–242

Chitarra W, Pugliese M, Gilardi G, Gullino ML, Garibaldi A (2013) Effect of silicates and electrical conductivity on Fusarium wilt of hydroponically grown lettuce. Commun Agric Appl Biol Sci 78:555–557

Dallagnol LJ, Rodrigues FA, Mielli MVB (2013) Silicon improves the emergence and sanity of rice seedlings obtained from seeds infected with *Bipolaris oryzae*. Trop Plant Pathol 38:478–484

Dallagnol LJ, Rodrigues FA, Mielli MVB, Ma JF (2014) Rice grain resistance to brown spot and yield are increased by silicon. Trop Plant Pathol 39:56–63

Dannon EA, Wydra K (2004) Interaction between silicon amendment, bacterial wilt development and phenotype of *Ralstonia solanacearum* in tomato genotypes. Physiol Mol Plant Path 64:233–243

Datnoff LE, Rodrigues FA, Seebold KW (2007) Silicon and plant disease. In: Datnoff LE, Elmer WH, Huber DM (eds) Mineral nutrition and plant disease. The American Phytopathological Society Press, St Paul, pp 233–246

Fortunato AA, Rodrigues FA, Baroni JCP, Soares GCB, Rodriguez MAD, Pereira OL (2012) Silicon suppresses Fusarium wilt development in banana plants. J Phytopathol 160:674–679

French-Monar RD, Rodrigues FA, Korndörfer GH, Datnoff LE (2010) Silicon suppresses Phytophthora blight development on bell pepper. J Phytopathol 158:554–560

Guérin V, Lebreton A, Cogliati EE, Hartley SE, Belzile F, Menzies JG, Bélanger RR (2014) A zoospore inoculation method with *Phytophthora sojae* to assess the prophylactic role of silicon on soybean cultivars. Plant Dis 98:1632–1638

Heine G, Tikum G, Horst WJ (2007) The effect of silicon on the infection by and spread of *Pythium aphanidermatum* in single roots of tomato and bitter gourd. J Exp Bot 58:569–577

Henriet C, Draye X, Oppitz L, Swennen R, Delvaux B (2006) Effects, distribution and uptake of silicon in banana (*Musa* spp.) under optimal conditions. Plant Soil 287:359–374

Huang CH, Roberts PD, Datnoff LE (2011) Silicon suppresses Fusarium crown and root rot of tomato. J Phytopathol 159:546–554

Kiirika LM, Stahl F, Wydra K (2013) Phenotypic and molecular characterization of resistance induction by single and combined application of chitosan and silicon in tomato against *Ralstonia solanacearum*. Physiol Mol Plant Pathol 81:1–12

Korndörfer GH, Datnoff LE, Corrêa GF (1999) Influence of silicon on grain discoloration and upland rice grown on four savanna soils from Brazil. J Plant Nutr 22:93–102

Lee JS, Seo ST, Wang TC, Jang HL, Pae DH, Engle LM (2004) Effect of potassium silicate amendment in nutrient solutions to suppress Phtophthora blight (*Phytophtora capsici*) in pepper (*Capsicum annuum*). World Veg Center 42:741

Liu L, Guo Y, Bi Y, Li M, Zhao J, Zhao H (2009) Inhabited mechanisms of silicon compounds against Fusarium rot (*Fusarium* spp.) of postharvest Chinese cantaloupe. J Food Process Preserv 33:187–202

McDonagh D, Hunter A (2010) Effect of silicon application on *Lolium perenne* development and Fusarium control. Paper presented at the XXVIII international horticultural congress on science and horticulture for people (IHC2010), Lisbon, Portugal, pp 195–201, Aug 2010

Menzies JG, Ehret DL, Chérif M, Bélanger RR (2001) Plant-related silicon research in Canada. In: Datnoff LE, Snyder GH, Korndörfer GH (eds) Silicon in agriculture. Elsevier Science B. V, Amsterdam, pp 323–341

Miyaki Y, Takahashi E (1983) Effect of silicon on the growth of solution-cultured cucumber plants. Jpn J Soil Sci Plant Nutr 29:71–83

Najihah NI, Hanafi MM, Idris AS, Hakim MA (2015) Silicon treatment in oil palms confers resistance to basal stem rot disease caused by *Ganoderma boninense*. Crop Prot 67:151–159

North Carolina State University (1997) Effect of soluble silica on brown patch and dollar spot of creeping bentgrass. North Carolina Turfgrass, Raleigh, Aug/Sept, pp 34–36

Oliveira RM, Ribeiro RCF, Xavier AA, Pimenta L, Korndörfer GH (2012) Effect of calcium and magnesium silicate on *Meloidogyne javanica* reproduction and development of banana Prata-Anã seedlings. Rev Bras Frutic 34:409–415

Prabhu AS, Barbosa Filho MP, Datnoff LE, Snyder GH, Berni RF, Rodrigues FA, Dallagnol LJ (2012) Silicon reduces brown spot severity and grain discoloration on several rice genotypes. Trop Plant Pathol 37:409–414

Rodgers-Gray BS, Shaw MW (2000) Substantial reductions in winter wheat diseases caused by addition of rice straw but not manure to soil. Plant Pathol 49:590–599

Rodgers-Gray BS, Shaw MW (2004) Effects of straw and silicon soil amendments on some foliar and stem-base diseases in pot-grown winter wheat. Plant Pathol 53:733–740

Rodrigues FA, Silva OA, Juliatti FC, Korndörfer GH, Corrêa GF, Peixoto JR (1996) Effect of calcium silicate on the development of *Fusarium oxysporum* f. sp. *lycopersici*, races 1 and 2, on tomato. Fitopatol Bras 21:393

Rondeau E (2001) Effect of potassium silicates on disease tolerance of bentgrass. Seminaire de fin d'etudes, Centre de Recherche en Horticulture, Université Laval, Québec

Saigusa M, Onozawa K, Watanabe H, Shibuya K (2000) Effects of porous hydrate calcium silicate on the wear resistance, insect resistance, and disease tolerance of turf grass "Miyako". Grassl Sci 45:416–420

Santos GR, Castro Neto MD, Carvalho ARS, Fidelis RR, Afférri FS (2010) Fontes e doses de silício na severidade do crestamentogomoso e produtividade da melancia. Biosci J 26:266–272

Schmidt RE, Zhang X, Chalmers DR (1999) Response of photosynthesis and superoxide dismutase to silica applied to creeping bentgrass grown under two fertility levels. J Plant Nutr 22:1763–1773

Seebold KW, Datnoff LE, Correa-Victoria FJ, Kucharek TA, Snyder GH (2000) Effect of silicon rate and host resistance on balst, scald and yield of upland rice. Plant Dis 84:871–876

Silva RV, Oliveira RDL, Nascimento KJT, Rodrigues FA (2010) Biochemical responses of coffee resistance against *Meloidogyne exigua* mediated by silicon. Plant Pathol 59:586–593

Sun XH, Sun YJ, Zhang CS, Song ZH, Chen J, Bai JK, Cui YY, Zhang C (1994) The mechanism of corn stalk rot control by application of potassic and siliceous fertilizers. Acta Phytophylacica Sin 21:102–108

Swain BN, Prasad JS (1988) Influence of silica content in the roots of rice varieties on the resistance to root-knot nematode. Indian J Nematol 18:360–361

Uriarte RF, Shew HD, Bowman DC (2004) Effect of soluble silica on brown patch and dollar spot of creeping bentgrass. J Plant Nutr 27:325–339

Vermeire ML, Kablan L, Dorel M, Delvaux B, Risède JM, Legrève A (2011) Protective role of silicon in the banana-*Cylindrocladium spathiphylli* pathosystem. Eur J Plant Pathol 131:621–630

Winslow MD (1992) Silicon, disease resistance, and yield of rice genotypes under upland cultural conditions. Crop Sci 32:1208–1213

Zhang Q, Fry J, Lowe K, Tisserat N (2006) Evaluation of calcium silicate for brown patch and dollar spot suppression on turf grasses. Crop Sci 46:1635–1643

Chapter 4
Silicon Control of Foliar Diseases in Monocots and Dicots

Fabrício A. Rodrigues, Leandro José Dallagnol, Henrique Silva Silveira Duarte, and Lawrence E. Datnoff

Abstract One of the most notable effects of silicon on plants is a reduction in the severity of multiple plant diseases caused by pathogenic organisms. For instance, the severity of several rice diseases, such as bacterial blight, brown spot, grain discoloration, leaf scald, leaf and panicle blast, stem rot and sheath blight were suppressed by applying silicon. The reduction in symptom expression is believed to be due to silicon's effects on multiple defense mechanisms in rice that govern the latent period, lesion size, lesion number, and inoculum production. Although foliar-applied silicon is effective in reducing many foliar diseases, applying silicon to the roots is more effective because it mediates the plant's defense responses to both foliar and root infections. Applications of silicon can perform as well as fungicides for suppressing plant diseases as well as enhance the resistance of susceptible cultivars to the same level as those that have complete genetic resistance. The application of silicon has value for inclusion in an integrated disease management strategy since this often overlooked, quasi-essential element clearly has the potential to reduced plant disease epidemics.

F.A. Rodrigues (✉)
Departamento de Fitopatologia, Universidade Federal de Viçosa,
Viçosa, Minas Gerais Cep 36570-900, Brazil
e-mail: fabricio.plantpathologist@gmail.com; fabricio@ufv.br

L.J. Dallagnol
Departamento de Fitossanidade, Universidade Federal de Pelotas,
Pelotas, Rio Grande do Sul Cep 96010-900, Brazil

H.S.S. Duarte
Departamento de Proteção de Plantas, SCA-DFF, Universidade Federal do Paraná
Curitiba, Paraná Cep 80035-050, Brazil

L.E. Datnoff
Department of Plant Pathology and Crop Physiology, Louisiana State University Agricultural Center, 302 Life Sciences, Baton Rouge, LA 70803, USA
e-mail: LDatnoff@agcenter.lsu.edu

F.A. Rodrigues, L.E. Datnoff (eds.), *Silicon and Plant Diseases*,
DOI 10.1007/978-3-319-22930-0_4

Introduction

A wide variety of monocot and dicot plants have benefited from silicon nutrition, whether direct or indirect, when exposed to different types and combinations of abiotic and biotic stresses. In addition to the many agronomic and horticultural benefits gained by maintaining adequate levels of this element in the soil and in plant tissue, silicon reduces the intensity of multiple plant diseases caused by biotrophic, hemibiotrophic and necrotrophic pathogens in many crops of great economic importance.

This chapter provides an overview of the potential effects of silicon on foliar diseases caused by fungi, bacteria and viruses in monocots and dicots. The effects of silicon on disease intensity in different host-parasite interactions are summarized in Table 4.1.

Table 4.1 Effect of silicon on some host-pathogen interactions

Hosts	Diseases	Pathogens	Effects[a]	References
Monocots				
Asparagus	Stem blight	*Phomopsis aspargi*	⊕	Lu et al. (2008)
Banana	Black Sigatoka	*Mycosphaerella fijiensis*	⊕	Kablan et al. (2012)
Barley	Black point	*Alternaria* spp.	⊕	Kunoh and Ishizaki (1975)
	Powdery mildew	*Blumeria graminis* f. sp. *hordei*	⊕	Leusch and Buchenauer (1989), Jiang (1993), Wiese et al. (2005)
Bermudagrass	Leaf spot	*Bipolaris cynodontis*	⊕	Datnoff et al. (2005)
Corn	Corn smut	*Ustilago maydis*	⊕	Tamimi and Hunter (1970)
Kentucky bluegrass	Powdery mildew	*Sphaerotheca fuliginea*	⊕	Hamel and Heckman (1999)
Pearl millet	Down mildew	*Sclerospora graminicola*	⊕	Deepak et al. (2008)
Perennial ryegrass	Gray leaf spot	*Pyricularia oryzae*	⊕	Nanayakkara et al. (2008, 2009)
Rice	Leaf and panicle blast	*Pyricularia oryzae*	⊕	Suzuki (1935), Volk et al. (1958), Datnoff et al. (1991), Seebold et al. (2000, 2001) Berni and Prabhu (2003), Nakata et al. (2008), Santos et al. (2009), Sun et al. (2010), Santos et al. (2011), Abed-Ashtiani et al. (2012), Cacique et al. (2012, 2013), Junior and Bonaldo (2013)

(continued)

Table 4.1 (continued)

Hosts	Diseases	Pathogens	Effects[a]	References
	Brown spot	*Bipolaris oryzae*	⊕	Takahashi (1967), Nanda and Gangopadhyay (1984), Yamauchi and Winslow (1987), Datnoff et al. (1990, 1991, 1992), Zanão Junior et al. (2009), Rezende et al. (2009), Santos et al. (2011), Prabhu et al. (2012), Dallagnol et al. (2009, 2013, 2014)
	Sheath blight	*Rhizoctonia solani*	⊕	Mathai et al. (1977), Datnoff et al. (1990), Winslow (1992), Rodrigues et al. (2001)
	Leaf scald	*Monographela albescens*	⊕	Yamauchi and Winslow (1989), Winslow (1992), Seebold et al. (2000), Tatagiba et al. (2014)
	Stem rot	*Magnaporthe salvinii*	⊕	Elawad and Green (1979)
	Grain discoloration	Many fungal species	⊕	Yamauchi and Winslow (1989), Winslow (1992), Korndörfer et al. (1999), Seebold et al. (2000), Prabhu et al. (2012)
	Bacterial leaf blight	*Xanthomonas oryzae* pv. *oryzae*	⊕	Chang et al. (2002)
Rye	Powdery mildew	*Erysiphe graminis*	⊕	Leusch and Buchenauer (1988)
Sorghum	Anthracnose	*Colletotrichum graminicola*	⊕	Narwal (1973), Resende et al. (2009, 2013)
St. Augustine grass	Gray leaf spot	*Pyricularia oryzae*	⊕	Datnoff and Nagata (1999), Brecht et al. (2007)
Sugarcane	Rust	*Puccinia melanocephala*	∅	Raid et al. (1992)
	Rust	*Puccinia melanocephala*	⊕	Naidoo et al. (2008), Camargo et al. (2013)
	Ring spot	*Leptosphaeria sacchari*	⊕	Raid et al. (1992)

(continued)

Table 4.1 (continued)

Hosts	Diseases	Pathogens	Effects[a]	References
Wheat	Powdery mildew	*Blumeria graminis* f. sp. *graminis*	⊕	Germar (1934), Leusch and Buchenauer (1989), Rodgers-Gray and Shaw (2000, 2004), Bélanger et al. (2003), Rémus-Borel et al. (2005), Guével et al. (2007), Côté-Beaulieua et al. (2009), Curtis et al. (2012)
	Septoria leaf blotch	*Septoria nodorum*	⊕	Rodgers-Gray and Shaw (2000, 2004)
	Leaf blast	*Pyricularia oryzae*	⊕	Xavier Filha et al. (2011), Pagani et al. (2014)
	Leaf rust	*Puccinia triticina*	⊗	Wordell Filho et al. (2013)
	Yellow spot	*Drechslera tritici-repentis*	⊗	Wordell Filho et al. (2013)
	Eyespot	*Oculimacula yallundae*	⊕	Rodgers-Gray and Shaw (2004)
	Rusts	*Puccinia* spp.	∅	Rodgers-Gray and Shaw (2004)
	Bacterial leaf streak	*Xanthomonas translucens* pv. *undulosa*	⊕	Silva et al. (2010)
	Spot blotch	*Bipolaris sorokiniana*	⊕	Domiciano et al. (2010a, b, 2013), Zanão Junior (2010)
Dicots				
Bean	Anthracnose	*Colletotrichum lindemuthianum*	⊕	Moraes et al. (2006, 2009), Polanco et al. (2012, 2014)
	Angular leaf spot	*Pseudocercospora griseola*	⊕	Rodrigues et al. (2010)
Coffee	Coffee leaf rust	*Hemileia vastatrix*	∅	Carré-Missio et al. (2009), Lopes et al. (2013a, 2014a)
	Coffee leaf rust	*Hemileia vastatrix*	⊕	Carré-Missio et al. (2012a, b, 2014), Pereira et al. (2009)
	Brown eye spot	*Cercospora coffeicola*	⊕	Pozza et al. (2004), Botelho et al. (2005)
	Brown eye spot	*Cercospora coffeicola*	∅	Lopes et al. (2013b)

(continued)

Table 4.1 (continued)

Hosts	Diseases	Pathogens	Effects[a]	References
Cotton	Ramularia leaf spot	*Ramularia areola*	∅	Aquino et al. (2008)
	Areolate mildew	*Ramularia gossypii*	⊕	Curvelo et al. (2013)
	Tropical rust	*Phakopsora gossypii*	⊕	Guerra et al. (2013a)
	Ramulosis	*Colletotrichum gossypii* var. *cephalosporioides*	⊕	Guerra et al. (2013b)
Cucumber	Powdery mildew	*Podosphaera xanthii*	⊕	Wagner(1940), Miyaki and Takahashi (1983), Adatia and Besford (1986), Menzies et al. (1991, 1992), Liang et al. (2005)
	Anthracnose	*Colletotrichum orbiculare*	⊕	Kanto (2002)
	Leaf spot	*Corynespora citrullina*	⊕	Kanto (2002)
	Gray mold rot Black rot	*Botrytis cinerea* *Didymella bryoniae*	⊕ ⊕	O'Neill (1991), Voogt (2001)
Eucalyptus	Powdery mildew	*Oidium eucalypti*	⊕	Schultz et al. (2012)
Gerbera daisy	Powdery mildew	*Podosphaera fusca*	⊕	Moyer et al. (2008)
Grape	Powdery mildew	*Uncinula necator*	⊕[b]	Bowen et al. (1992)
	Powdery mildew	*Uncinula necator*	∅	Blaich and Grundhöfer (1998)
Lettuce	Downy mildew	*Bremia lactucae*	⊕	Garibaldi et al. (2012)
Melon	Powdery mildew	*Podosphaera xanthii*	⊕	Dallagnol et al. (2012)
	Pink rot	*Trichothecium roseum*	⊕	Bi et al. (2006)
	Alternaria	*Alternaria alternata*	⊕	Bi et al. (2006)
	Fusarium	*Fusarium semitectum*	⊕	Bi et al. (2006)
Morning glory	Anthracnose	*Colletotrichum gloeosporioides*	⊕	Kunoh and Ishizaki (1975)
Muskmelon	Powdery mildew	*Podosphaera xanthii*	⊕	Menzies et al. (1992)
	Bacterial fruit blotch	*Acidovorax citrulli*	⊕	Conceição et al. (2014)
	Pink rot	*Trichothecium roseum*	⊕	Li et al. (2012)
Paper daisies	Anthracnose	*Colletotrichum gloeosporioides*	⊕	Muir et al. (2001)

(continued)

Table 4.1 (continued)

Hosts	Diseases	Pathogens	Effects[a]	References
Pea	Leaf spot	*Mycosphaerella pinodes*	⊕	Dann and Muir (2002)
Peach	Brown rot	*Monilinia fructicola*	⊕	Yang et al. (2010)
Potato	Late blight	*Phytophthora infestans*	∅	Duarte et al. (2008), Soratto et al. (2012)
Pumpkin	Powdery mildew	*Sphaerotheca xanthii*	⊕	Heckman et al. (2003)
Rose	Powdery mildew	*Sphaerotheca pannosa*	⊕	Voogt (1992), Shetty et al. (2012)
Soybean	Stem canker	*Diaporthe phaseolorum* f. sp. *meridionalis*	⊕	Juliatti et al. (1996), Grothge-Lima et al. (2001)
	Asian soybean rust	*Phakopsora pachyrhizi*	⊕	Pereira et al. (2009), Rodrigues et al. (2009), Lemes et al. (2011), Arsenault-Labrecque et al. (2012), Cruz et al. (2012, 2013, 2014b)
	Asian soybean rust	*Phakopsora pachyrhizi*	∅	Duarte et al. (2009)
Strawberry	Powdery mildew	*Sphaerotheca macularis* f. sp. *macularis*	⊕	Voogt and Sonneveld (2001), Kanto (2002), Kanto et al. (2004, 2006, 2007)
	Pestalotia leaf spot	*Pestalotia longisetula*	⊕	Carré-Missio et al. (2010)
	Gray mold	*Botrytis cinerea*	∅	Lopes et al. (2014a)
	Anthracnose fruit rot	*Colletotrichum acutatum*	⊕	Igarashi (2008)
Tobacco	BdMV	*Belladonna mottle virus*	⊗	Bengsch et al. (1989)
	TMV	*Tobacco mosaic virus*	∅	Zellner et al. (2011)
	TRSV	*Tobacco ringspot virus*	⊕	Zellner et al. (2011)
Tomato	Powdery mildew	*Oidiopsis sicula*	⊕[b]	Menzies et al. (2001)
	Powdery mildew	*Oidiopsis sicula*	∅	Yanar et al. (2011)
		Oidium neolicopersici	⊕	Garibaldi et al. (2011)
	Late blight	*Phytophthora infestans*	∅	Duarte et al. (2007)
	Bacterial speck	*Pseudomonas syringae* pv. *tomato*	⊕	Andrade et al. (2013)
Yellow passion fruit	Bacterial spot	*Xanthomonas axonopodis* pv. *passiflorae*	⊕	Brancaglione et al. (2009)

(continued)

Table 4.1 (continued)

Hosts	Diseases	Pathogens	Effects[a]	References
Zucchini squash	Powdery mildew	*Podosphaera xanthii*	⊕	Menzies et al. (1992), Savvas et al. (2009), Tesfagiorgis and Laing (2011), Tesfagiorgis et al. (2014)

[a]Silicon can decrease (⊕), increase (⊗) or has no effect (∅) on disease intensity
[b]Silicon decreases disease intensity if foliarly applied, but has no effect on disease when added to the nutrient solution

Fungal Diseases

Rice Silicon application to paddy soils, which enhances rice's resistance to blast (*Pyricularia oryzae*), began in Japan (Suzuki 1935). Volk et al. (1958) later proved that the increase in rice resistance to blast occurred through a reduction in the number of lesions as foliar silicon concentrations increased. Seebold et al. (2000) reported that the application of silicon at one experimental site in eastern Colombia reduced the severity of rice blast in a partially resistant cultivar to levels observed in a resistant cultivar that had not been amended with silicon. In another study, Seebold et al. (2001) evaluated four rice cultivars with different levels of susceptibility to *P. oryzae* and observed that with increasing silicon soil concentrations, the incubation period was lengthened; and the numbers of sporulating lesions, lesion size, rate of lesion expansion, diseased leaf area and number of conidia produced per lesion were significantly reduced in the most susceptible cultivars, M201, Rosemont and Lemont (Figs. 4.1, 4.2 and 4.3). According to these authors, the net effect of silicon on these components of host resistance was an overall reduction in the production of conidia on plants infected with *P. oryzae*, which slowed the epidemic rate of the blast. *Lsi*1 mutant rice plants deficient in active silicon uptake had very low silicon shoot concentrations and, as a consequence, were more susceptible to blast development under field conditions (Nakata et al. 2008). A negative correlation between silicon shoot concentration and rice's susceptibility to blast was reported for many cultivars with varying silicon concentrations (Kozaka 1965; Ou 1985; Abed-Ashtiani et al. 2012). Rabindra et al. (1981) found that the silicon concentration in leaves and panicles varied among four rice cultivars grown under similar climatic conditions and that those cultivars that accumulated more silicon in their shoots showed less incidence of leaf and panicle blast. In that study, the accumulation of silicon in rice shoots appeared to be dependent on genotype; however, environmental conditions at the site of plant growth also may have played a pivotal role. Rice plants grown under elevated CO_2 concentrations were more susceptible to leaf blast as indicated by an increase in the area under a disease progress curve and the number of lesions formed in those plants compared with plants grown under ambient CO_2 concentration conditions, regardless of whether the experiment was conducted under controlled or field conditions (Kobayashi et al. 2006; Gória et al. 2013).

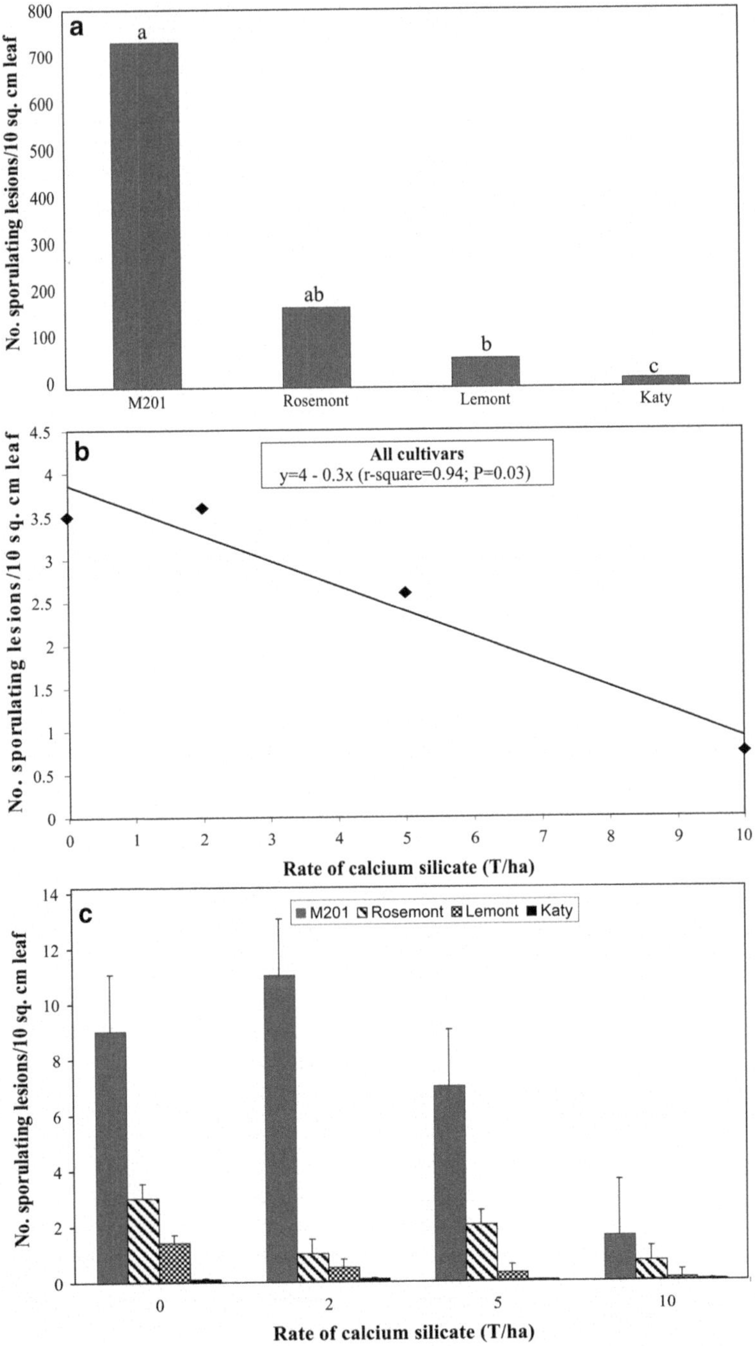

Fig. 4.1 Effects of calcium silicate on the number of sporulating lesions of blast (relative infection efficiency) for rice cultivars M201 (susceptible), Rosemont (moderately susceptible), Lemont (moderately susceptible), and Katy (resistant). (**a**) Mean number of sporulating lesions per square millimeter of leaf for each cultivar averaged across calcium silicate rates. Bars with the same letter

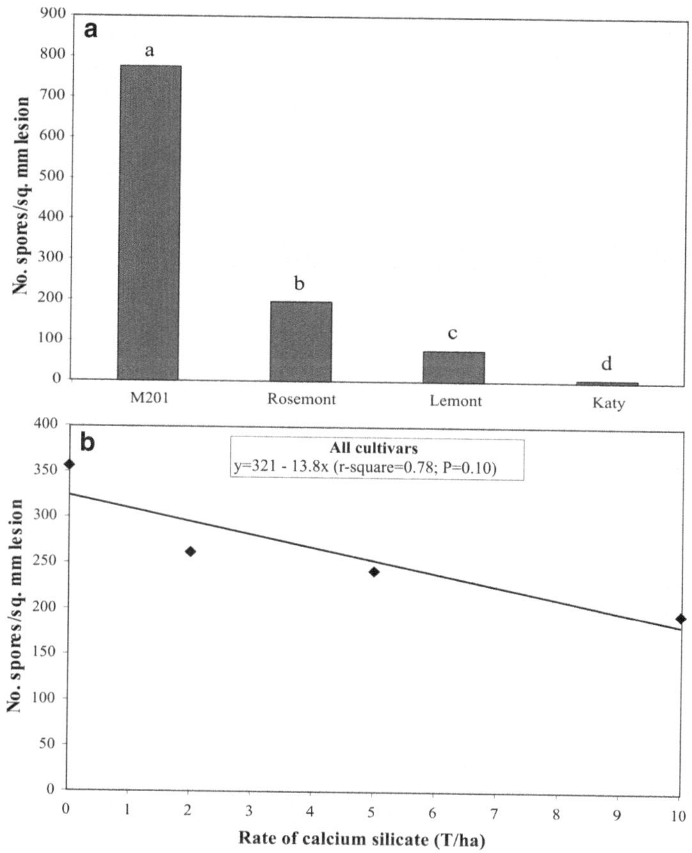

Fig. 4.2 Effects of calcium silicate on the sporulation of blast lesions for cultivars M201 (susceptible), Rosemont (moderately susceptible), Lemont (moderately susceptible), and Katy (resistant). (**a**) The number of spores per square millimeter of lesions for each cultivar averaged across calcium silicate rates. Bars with the same letter do not differ significantly at $P = 0.05$, as determined by Fisher's protected LSD test performed on log-transformed values. (**b**) The relationship between the number of spores per square millimeter of lesions and calcium silicate rate averaged across means for all cultivars (Reproduced from Datnoff and Rodrigues (2005))

Interestingly, under artificial inoculation, panicle blast severity was not affected by CO_2 enrichment (Kobayashi et al. 2006). However, CO_2 enrichment did result in lower silicon accumulation in shoot tissue, which was correlated with an increase in rice susceptibility to blast (Kobayashi et al. 2006; Gória et al. 2013). Kobayashi et al. (2006) highlighted that a simple increase in CO_2 concentration did not neces-

Fig. 4.1 (continued) in A do not differ significantly at $P = 0.05$, as determined by Fisher's protected LSD test performed on log-transformed values. (**b**) The relationship between the number of sporulating lesions per square millimeter of leaf and calcium silicate rates averaged across means for all cultivars. (**c**) The number of sporulating lesions per square millimeter of leaf for each cultivar and calcium silicate rates. Bars represent standard errors of means (Reproduced from Datnoff and Rodrigues (2005))

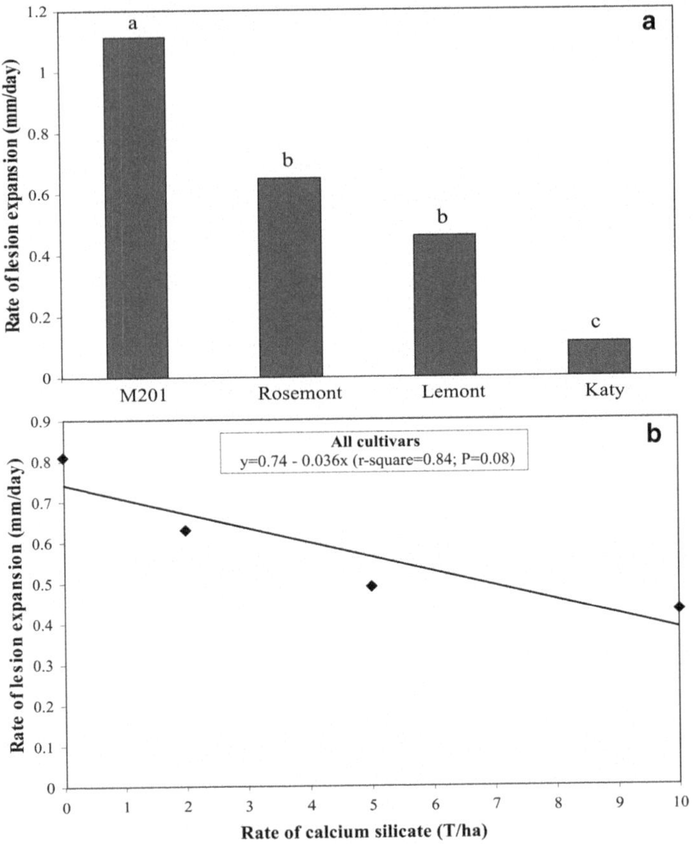

Fig. 4.3 Effects of calcium silicate on the daily rate of expansion of blast lesions for rice cultivars M201 (susceptible), Rosemont (moderately susceptible), Lemont (moderately susceptible), and Katy (resistant). (**a**) The rate of expansion of blast lesions for each cultivar averaged across calcium silicate rates. Bars with the same letter do not differ significantly at $P=0.05$, as determined by Fisher's protected LSD test. (**b**) The relationship between the rate of lesion expansion and calcium silicate rates averaged across means for all cultivars (Reproduced from Datnoff and Rodrigues (2005))

sarily result in an increase in rice susceptibility to blast. They also noted that in rice production years where a longer light duration and lower air humidity were recorded during rice development, silicon transport seemed to accelerate from the roots to the shoots. This action consequently raised the silicon concentration to its maximum tissue level and reduced blast development. However, other reports have shown that rice cultivars accumulating higher levels of silicon in shoots are not always more resistant to blast than cultivars accumulating lower levels of silicon when grown under the same environmental conditions (Kozaka 1965; Ou 1985; Winslow 1992). The magnitude of disease control in rice can be strongly affected by the silicon source and concentration. The incidence of rice blast was reduced by nearly 50 %

for rice plants amended with various inorganic and organic silicon sources relative to plants not receiving silicon (Aleshin et al. 1987). In Nigeria, the application of sodium silicate to upland rice grown in a silicon-depleted soil decreased panicle blast severity on three cultivars by approximately 40 % (Yamauchi and Winslow 1989). Winslow (1992) reported that the addition of sodium metasilicate to silicon-deficient soils in Nigeria greatly reduced panicle blast severity on eight different rice genotypes by over 50 %. In southern Florida, the amendment of 5, 10 and 15 t/ha of calcium silicate slag in a silicon-deficient Histosol linearly and curvilinearly reduced panicle blast development in rice (Datnoff et al. 1991). Additional studies conducted with calcium silicate slag revealed that finely ground grades were more effective than more coarsely ground grades in reducing the intensity of panicle blast, and the use of finely ground grades of slag was correlated with both higher silicon concentrations in rice shoots and increased yields (Datnoff et al. 1992). Datnoff and Snyder (1994) demonstrated that reductions in the severity of panicle blast brought about by the application of 0.4 t of elemental silicon/ha did not differ significantly from those achieved by applying a labeled amount of the fungicide benomyl. In their studies, disease severity was negatively correlated with silicon concentrations in shoots. Indeed, a single application of silicon had a significant residual effect on the control of leaf and panicle blast in the next rice growing season. According to Datnoff et al. (1991), the application of calcium silicate slag in 1987 reduced panicle blast by 31 % and brown spot by 15 % over the control (Fig. 4.4). In 1988, panicle blast and brown spot were reduced by 17 % and 32 % over the control, respectively (Figs. 4.5 and 4.6). In 1988, brown spot severity at the highest calcium silicate slag concentration decreased by 15 %, 18 % and 17 % over the control for residual 1987 slag effects on the 1988 rice crop, 1988 slag applications,

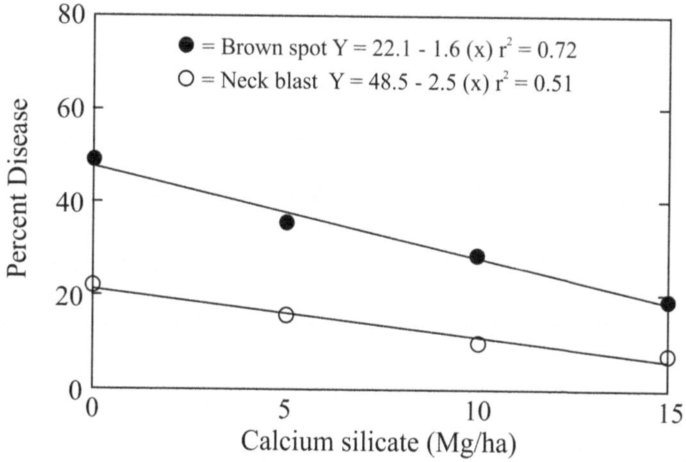

Fig. 4.4 Relationship of brown spot severity and panicle (neck) blast incidence to rates of calcium silicate slag in 1987 (Reproduced from Datnoff et al. (1997), with permission from Elsevier)

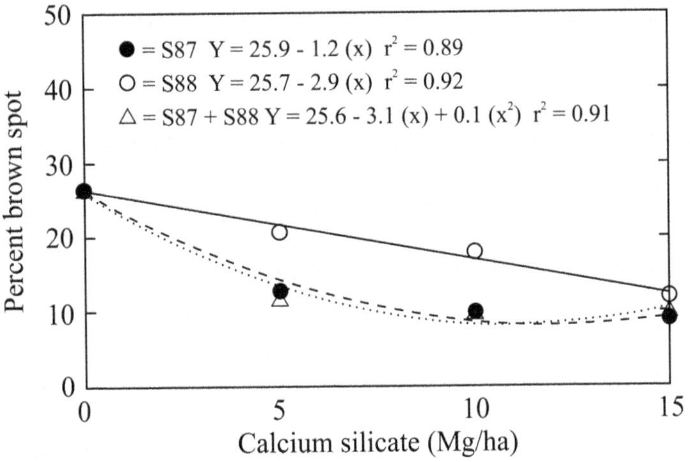

Fig. 4.5 Relationship of brown spot severity to calcium silicate slag quantities in 1988. S87 or S88 = slag applied at given treatment rates only in 1987 or 1988; S87 + S88 = 5 Mg/ha of slag applied in 1988 to each of the residual plots receiving the 1987 slag treatments (Reproduced from Datnoff et al. (1997), with permission from Elsevier)

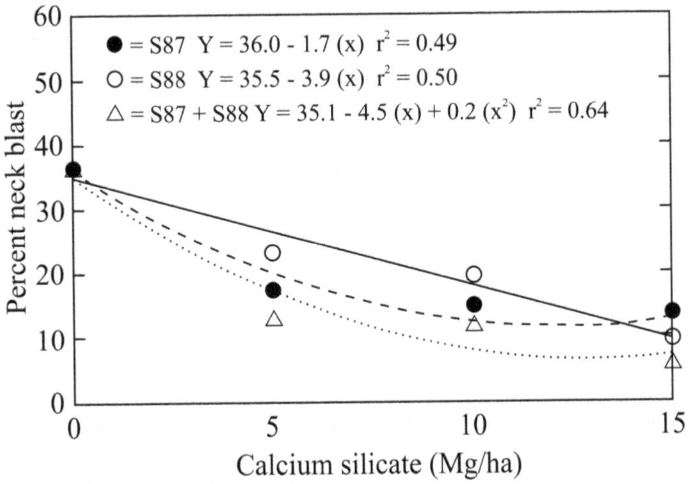

Fig. 4.6 Relationship of panicle (neck) blast incidence to rates of calcium silicate slag in 1988. S87 or S88 = slag applied at given treatment rates only in 1987 or 1988; S87 + S88 = 5 Mg/ha of slag applied in 1988 to each of the residual plots receiving the 1987 slag treatments (Reproduced from Datnoff et al. (1997), with permission from Elsevier)

and residual 1987 slag rates that received 5 t/ha of slag in 1988, respectively (Fig. 4.5).

Silicon also reduced the intensity of sheath blight (*Rhizoctonia solani*), even though there was no significant difference between the blight intensity levels found in plants treated with low and high silicon concentrations (Mathai et al. 1977). Winslow (1992) reported that silicon only reduced the severity of sheath blight in irrigated *indica* rice genotypes but not in *japonica* upland rice or intermediate genotypes. However, Rodrigues et al. (2001) showed that silicon decreased the severity of sheath blight in both tropical *japonicas* and an *indica* rice cultivar, indicating that the enhanced rice resistance provided by silicon was not limited to *indica* cultivars. In fact, the authors noted that silicon reduced the intensity of sheath blight in two susceptible (Lemont and Labelle) and two moderately susceptible (Drew and Kaybonnet) rice cultivars down to the same levels of intensity observed in two cultivars (Jasmine and LSBR-5) with high partial resistance to sheath blight that had not been amended with silicon. In a silicon-deficient typic acrustox red yellow latosol from Brazil, the application of silicon in six rice cultivars significantly reduced the total number of sheath blight lesions on sheaths, the total area under the relative lesion extension progress curve, the severity of sheath blight, and the highest relative lesion height on the main tiller by 37 %, 40 %, 52 % and 24 %, respectively, as the soil concentration of silicon increased from 0 to 8 t/ha (Rodrigues et al. 2003b). Furthermore, the effect of silicon in reducing rice susceptibility to sheath blight was independent of rice growth stages, suggesting that silicon may increase rice resistance to this disease at all developmental stages of a plant's growth (Rodrigues et al. 2003a).

Leaf scald (*Monographella albescens*) severity was also reduced as much as 42 % with increasing rates of silicon amendments to soil (Seebold et al. 2000). Tatagiba et al. (2014) reported that foliar silicon concentrations significantly increased in rice plants supplied with silicon (4.8 dag/kg) compared with those that did not receive the treatment (0.9 dag/kg). Higher foliar silicon concentrations were reported to significantly reduce the expansion of leaf scald lesions by 21 %, 15 % and 18 % at 72, 96 and 120 h after inoculation, respectively, compared with the untreated plants.

Prabhu et al. (2012) reported a negative correlation between brown spot (*Bipolaris oryzae*) severity and silicon concentration. The area under the brown spot progress curve was reduced up to 75 % by silicon under greenhouse conditions (Dallagnol et al. 2009). This reduction in brown spot development was due to silicon's effect on a number of components of host resistance, including an increase in the incubation period and a reduction in the relative efficiency of infection, the rate of lesion expansion and the final lesion size. Furthermore, the study revealed that in order for silicon to decrease brown spot symptoms, foliar silicon concentrations needed to reach a certain minimum value. This was based on an experiment where plants of the rice mutant *lsi*1, which exhibit low active silicon efflux, accumulated a lower level of silicon in leaf tissue and produced a greater area under the disease progress curve in comparison with plants from the wild cultivar Oochikara (Dallagnol et al. 2009). Cacique et al. (2012) studied the effects of silicon (0 or 2 mmol) and manganese (0.5, 2.5, and 10 mmol) concentrations, and their interaction, on rice resistance to blast. Silicon concentrations were significantly higher in the leaf tissue of plants supplied with this element than those that were untreated, regardless of manganese concentrations. No significant differences in silicon concentrations were observed for changing manganese concentrations for plants treated or not treated with silicon. The incubation period of blast increased for the plants supplied with silicon. Manganese concentrations had no effect on the incubation period, regardless of whether silicon was supplied or not. The number of lesions per cm^2 of leaf area significantly decreased in the presence of silicon, regardless of manganese concentrations. In the presence of silicon, lesion size and the area under the blast progress curve were significantly reduced, regardless of manganese concentrations. However, in the absence of silicon, the values for lesion size and area under the blast progress curve were significantly lower at the manganese concentration of 10 μmol compared with 0.5 μmol. Overall, the authors showed the potential of silicon to decrease rice blast development, regardless of the presence or absence of manganese in leaves.

Small Grains Powdery mildew (*Blumeria graminis* f. sp. *tritici*) on wheat can be efficiently controlled by silicon (Germar 1934; Rodgers-Gray and Shaw 2000, 2004). In one study, the first signs of infection by *B. graminis* f. sp. *tritici* were observed at 4 days after the inoculation of plants not amended with silicon; and the disease developed rapidly thereafter, reaching severities of up to 40 % after 5 weeks (Bélanger et al. 2003). By contrast, colonies of *B. graminis* f. sp. *tritici* remained low even after 5 weeks in plants amended with silicon, and severities stayed below 5 %, indicating very limited fungal colonization of leaf tissue. Guével et al. (2007) also reported reductions as high as 80 % in powdery mildew severity on wheat leaves when silicon was applied via the roots. A reduction in spot blotch (*Bipolaris sorokiniana*) severity up to 28 % was reported for wheat plants grown in two silicon-deficient Latosols (Yellow Latosol and Red Latosol) amended with calcium silicate (Zanão Júnior et al. 2010). According to the study's authors, no significant difference was detected between the two soils types; but when silicon was applied, the incubation period increased and the area under the spot blotch disease progress curve was reduced. In another study of silicon's effect on spot blotch, the highest foliar silicon concentration reduced the area under disease progress curve by 59 % (Domiciano et al. 2010b). The effect of silicon on spot blotch development included an increase in the incubation period, a decrease in the number of lesions per cm^2 of leaf area and a decrease in disease severity; however, silicon had no significant effect on lesion size (Domiciano et al. 2010b). The highest foliar silicon concentration found in the flag leaves of wheat plants supplied with silicon also reduced the severity of spot blotch (Domiciano et al. 2010a). An analysis of how silicon affects the infectious process of *B. sorokiniana* on wheat leaves revealed that the number of brown (necrotic) epidermal cells and the frequency of infection sites with browning were significantly lower in leaves of silicon-amended plants. In addition, only a sparse network of hypha was found colonizing cells, indicating limited fungal growth within the tissue of silicon-amended plants (Domiciano et al. 2013). Wheat plants amended with silicon showed a 28 % increase in the incubation period for blast (*P. oryzae*) and reductions of up to 45 % and 31 % for the number of lesions per cm^2 of leaf area and area under the blast progress curve, respectively, when compared with non-amended plants (Xavier Filha et al. 2011). However, the authors reported that silicon had no significant effect on final disease severity. Wheat plants amended with silicon were less severely impacted by leaf blotch (*Parastagnospora* *nodorum*) under both field and greenhouse conditions (Rodgers-Gray and Shaw 2000, 2004). Silicon also reduced septoria leaf blotch (*Mycosphaerella graminicola*) and eyespot (*Oculimacula yallundae*). However, silicon's ability to suppress these diseases varied and was attributed to the type of growing substrate used in the experiments (Rodgers-Gray and Shaw 2004). Silicon also reduced the incidence of smut (*Ustilago maydis*) on corn, but only in association with phosphorus fertilization (Tamimi and Hunter 1970). Silicon amendments efficiently reduced the number of *B. graminis* f. sp. *hordei* colonies on barley leaves (Wiese et al. 2005). Similarly, soil amended with calcium silicate increased the foliar silicon concentration and reduced anthracnose (*Colletotrichum sublineolum*) severity in a susceptible cultivar of sorghum (Resende et al. 2009). For resistant sorghum cultivars, silicon did not have a significant influence on the measures of host resistance evaluated. However, for the susceptible cultivar, a negative correlation was found between foliar silicon concentrations and the area under the relative infection efficiency progress curve, the area under the anthracnose index progress curve, the percentage of necrotic leaf area and the final disease severity (Resende et al. 2009). In a field experiment, Resende et al. (2013) reported reductions of up to 42 % in the area under the anthracnose progress curve for plants from a susceptible cultivar grown in a silicon-deficient soil amended with calcium silicate. Furthermore, the residual effect of calcium silicate in the soil increased foliar silicon concentrations and yields, as well as reduced the intensity of anthracnose during the following growing season (Resende et al. 2013).

Cucurbits The application of soluble silicon via a recirculating nutrient solution reduced the severity of powdery mildew (*Podosphaera xanthii*) in cucumber (Miyaki and Takahashi 1983; Adatia and Besford 1986; Menzies et al. 1991). Contrary to these authors' findings, Schuerger and Hammer (2002) showed that silicon did not effectively reduce powdery mildew severity on plants from susceptible cucumber cultivars grown in nutrient solutions with potassium silicate concentrations of 100, 150 and 200 mg/L. According to Schuerger and Hammer, although the intensity of powdery mildew was slightly lowered in plants supplemented with silicon, disease suppression by silicon was not commercially useful because it failed to increase yield. Temperature was found to act in a synergistic manner with the level of silicon applied. Disease suppression by silicon was observed at 25 °C and 30 °C, but the magnitude of the response was significantly lower in comparison with those plants maintained at 20 °C. Foliar sprays of potassium silicate at and above 17 mM (1000 ppm) were effective in controlling powdery mildew (*P. xanthii*) on muskmelon and zucchini (Menzies et al. 1992). Tesfagiorgis and Laing (2011) also reported that applying silicon onto the leaves of zucchini plants reduced the severity of powdery mildew. However, as reported for cucumber, the efficacy of silicon was improved by increasing the application frequency. The best results were obtained when the foliar application frequency was increased to three times per week and when the runoff from the foliar silicon application reached the roots of the plants and was taken up (Tesfagiorgis and Laing 2011). The use of soluble silicon associated with potential biocontrol agents to control powdery mildew on zucchini was investigated under greenhouse and field conditions (Tesfagiorgis et al. 2014). The biocontrol agents were applied foliarly, while the potting mix containing the growing plants was saturated weekly with silicon. Under greenhouse conditions, biocontrol agents reduced powdery mildew severity by up to 90 %. Silicon alone reduced powdery mildew by as much as 35 % and improved the efficacy of most of the biocontrol agents (Tesfagiorgis et al. 2014). Under field conditions, powdery mildew reduction ranged from 10 % to 70 % when the biocontrol agents and silicon were applied; but the efficacy of the bacterial biocontrol agent *Serratia marcescens* and silicon diminished when the temperature increased above 25 °C (Tesfagiorgis et al. 2014). According to Tesfagiorgis et al., the level of powdery mildew suppression provided by each control option was promising under both greenhouse and field conditions, especially when they were combined. However, high levels of disease intensity reduced the efficacy of both the biocontrol agents, especially *S. marcescens* and silicon. This suggests that fungicides and/or resistant cultivars would need to be used in combination with silicon to help keep disease at a desirable level so that yields are not impacted. Soluble silicon (100 mM) significantly inhibited the *in vitro* mycelial growth of *Alternaria alternata, Fusarium semitectum* and *Trichothecium roseum* in Hami melons (*Cucumis melo* L. var. *inodorus* Jacq.) (Bi et al. 2006). The use of 100 mM silicon pre-inoculated with *T. roseum* resulted in a lower incidence of plant deterioration and reduced disease severity than treatments applied after inoculation (Bi et al. 2006).

Beans and Soybeans The relationships between soybean rust (*Phakopsora pachyrhizi*) severity and potassium silicate concentrations (ranging from 8 to 60 g/L) were linear and quadratic for solutions adjusted to pHs of 5.5 and 10.5, respectively (Rodrigues et al. 2009). According to Rodrigues et al., soybean rust severity at the highest potassium silicate concentration (pH 5.5) was 70 % lower than the control treatment (plants sprayed with a potassium hydroxide solution). Under greenhouse conditions, 40 g/L of potassium silicate reduced soybean rust severity and the number of pustules per cm^2 of leaf to a level comparable with reductions observed in plants treated with the fungicides epoxiconazole and pyraclostrobin (Rodrigues et al. 2009). In a more detailed study of the effect of silicon on soybean resistance to rust using both light and scanning electron microscopes, leaflets of plants supplied with silicon were found to have uredia that were smaller and fewer in number than those left untreated. Furthermore, reductions of 27 %, 23 % and 60 % occurred in the number of lesions per cm^2 of leaf area, closed uredia and open uredia, respectively, in the leaflets of plants amended with silicon (Cruz et al. 2012). In bean plants grown in a hydroponic culture amended with silicon, the incubation period increased by 14 %, while the area under the anthracnose progress curve and disease severity were reduced by 33 % and 34 %, respectively (Polanco et al. 2012). Cruz et al. (2014a) investigated the effect of silicon on bean resistance to infection by *C. lindemuthianum* at the microscopic level. According to these authors, disease severity decreased by 52 % in the leaves of plants supplied with silicon (4.4 %) compared with the leaves of untreated plants (8.5 %). Observations made using scanning electron microscopy revealed morphological changes in the veins of the leaves of bean plants not supplied with silicon compared with the leaves of plants supplied with silicon. X-ray microanalysis has revealed that mineral concentrations of potassium, silicon, and sulfur were higher in the leaves of plants supplied with silicon, which may have contributed to a decrease in anthracnose symptoms.

Perennial Crops Coffee plant roots were found to be inefficient in the uptake and translocation of silicon from a nutrient solution to the shoots. Therefore, no increase in the plant's resistance to coffee rust (*Hemileia vastatrix*) was observed (Carré-Missio et al. 2009). Similarly, a 3 year-long field experiment showed that soil amended with calcium silicate neither increased foliar silicon concentrations nor reduced the area under the rust progress curve in coffee plants (Lopes et al. 2013a). However, the foliar application of potassium silicate at pH 5.5 reduced coffee leaf rust development (Pereira et al. 2009). Furthermore, Carré-Missio et al. (2012b) reported significant reductions in coffee leaf rust severity, the sporulation intensity of *H. vastatrix* and the total number of pustules per cm^2 of leaf area due to foliar applications of potassium silicate. Carré-Missio et al. (2012a) also investigated possible local and systemic protections provided by potassium silicate sprayed on coffee leaves to reduce the symptoms of coffee leaf rust. In the first experiment, coffee plants with three pairs of leaves were sprayed with potassium silicate, epoxiconazole, acibenzolar-S-methyl (ASM) and distilled water following two methods: (1) spraying the 3rd pair of leaves from the apex and protecting the 2nd pair of leaves, or (2) spraying the two pairs of leaves on the left side of the plant and protecting the pair of leaves on the right side. After 24 h, the abaxial surface of the protected pair of leaves was inoculated with *H. vastatrix*. In the second experiment, the 3rd pair of leaves from the apex was sprayed with potassium silicate, ASM and distilled water and the 2nd pair of leaves was protected. At 1, 5, 15, 25 and 35 days after treatment applications, the abaxial surface of the 2nd (systemic protection) and 3rd (local protection) pairs of leaves for each treatment were inoculated with *H. vastatrix*. The potassium silicate sprayed on the 3rd pair of leaves or on the pair of leaves on the left side was ineffective in increasing silicon concentrations as well as in decreasing the intensity of sporulation, the total number of pustules per leaf and rust severity on the 2nd pair of leaves and the pair of leaves on the right side. This was in contrast to epoxiconazole and ASM, which affected the plant systemically. The potassium silicate sprayed on the 3rd pair of leaves, which acted non-systemically, also did not reduce the intensity of sporulation, the total number of pustules per leaf or rust severity compared to the local levels of protection. Therefore, potassium silicate can be used to reduce the intensity of coffee leaf rust preventively only at the point of surface leaf contact because no known silicon transporter genes are found in coffee leaf tissue (Carré-Missio et al. 2012a). Under field conditions with high incidence of coffee leaf rust, the application of potassium silicate alone or in combination with copper hydroxide was not very effective in reducing coffee leaf rust incidence or increasing yield (Lopes et al. 2013b). Contrary to previously reported greenhouse and field experiments (Carré-Missio et al. 2009, 2012a, b; Pereira et al. 2009; Lopes et al. 2013b), X-ray analysis revealed that coffee plants grown in soil amended with calcium silicate showed an increase in foliar silicon concentrations and a uniform abaxial leaf surface distribution for plants from Catuaí, Mundo Novo and Icatú cultivars (Pozza et al. 2004). However, this increase in foliar silicon concentrations was associated with a 63 % reduction in brown eye spot (*Cercospora coffeicola*) lesions and a 43 % reduction in total lesions per plant compared with those that were not amended; however, this only occurred in plants from the Catuaí cultivar (Pozza et al. 2004).

Banana plants supplied with silicon showed reduced black sigatoka (*Mycosphaerella fijiensis*) severity compared with plants not amended with this element (Kablan et al. 2012). In sugarcane, the reduction of brown rust (*Puccinia melanocephala*) severity through the use of silicon is controversial. Raid et al. (1992) observed that silicon had no influence on brown rust development in sugarcane; however, the brown rust severities observed in this 2-year study were 13 % or less, and it is possible that these disease severity values were not high enough to detect treatment differences. Foliar- and soil-applied silicon were tested to reduce sugarcane brown rust development in South Africa and resulted in reductions of up to 25 % in sugarcane plants supplied with silicon via soil (calcium silicate) and foliage (potassium silicate) (Naidoo et al. 2008). In Brazil, the silicon amendments to three different soil types (Quartzipsamment-RQ; Rhodic Hapludox-LV; Rhodic Acrudox-LVdf) resulted in reductions in the incidence of brown rust as silicon concentrations increased, but these results were dependent on soil type (Camargo et al. 2013). The application of silicon reduced the maximum rust incidence, as estimated by the b1 parameter of the monomolecular model, by 29 %, 41 % and 47 % for the RQ, LV and LVdf soil types, respectively (Camargo et al. 2013). Sugarcane plants grown in soil amended with silicon showed a significant reduction in ring spot (*Leptosphaeria sacchari*) severity by 67 % (Raid et al. 1992).

Turfgrass Powdery mildew (*Sphaerotheca fuliginea*) in Kentucky bluegrass (Hamel and Heckman 1999) and leaf spot (*Bipolaris cynodontis*) in bermudagrass (Datnoff et al. 2005) were reduced by an increase in silicon. Gray leaf spot (*P. oryzae*) development was reduced by silicon amendments of 19–78 % on several cultivars of St. Augustinegrass under greenhouse conditions (Datnoff and Nagata 1999); and in a subsequent study, Brecht and his colleagues (2007) demonstrated that lesion number was the only component of host plant resistance affected by silicon in the St. Augustinegrass-*P. oryzae* host-pathogen system. In perennial ryegrass, the incidence and severity of gray leaf spot also was significantly reduced as silicon concentrations increased in the soil and plant tissue (Nanayakkara et al. 2008, 2009).

Edible Vegetables In tomato plants grown in a hydroponic system with three levels of electrical conductivity (1.8–2 mS/cm (EC1), 3.9–4 mS/cm (EC2, 0.87 g/L NaCl) and 5–5.5 mS/cm (EC3, 1.74 g/L NaCl)), the addition of 100 mg/L of potassium silicate reduced powdery mildew (*Oidium neolycopersici*) incidence and severity, especially at the EC2 conductivity level (Garibaldi et al. 2011). According to these authors, the addition of sodium chloride to the nutrient solution also reduced the incidence and severity of powdery mildew. The addition of potassium silicate to the control nutrient solution also resulted in a similar or higher level of control over powdery mildew than the use of a nutrient solution with a higher conductivity but no silicon amendment (Garibaldi et al. 2011). However, considering that tomatoes accumulate very little silicon in their shoots, and foliar silicon concentrations were not determined in their study, the reduction of powdery mildew severity may not be attributed to silicon. Under field conditions, Yanar et al. showed that potassium silicate applied every 12 days was effective in reducing powdery mildew (*Leveillula taurica*) severity down to 6 % in comparison with the control treatment (78 %) (Yanar et al. 2011). In contrast, Duarte et al. found that, for tomatoes, potassium silicate application rates of 5 and 15 g/L were ineffective in reducing late blight (*Phytophthora infestans*) severity, while fungicides reduced disease severity up to 93 % (Duarte et al. 2007). In potatoes, the foliar application of silicon reduced the severity of late blight up to 35 % and increased both tuber yield and tuber dry matter content (Soratto et al. 2012). Asparagus plants amended with silicon made available through the roots increased silicon concentrations in both the roots and shoots, which in turn reduced the severity of stem blight (*Phomopsis asparagi*) in susceptible and partially resistant cultivars by up to 32 % (Lu et al. 2008). A susceptible asparagus cultivar amended with silicon reached a disease index similar to an asparagus cultivar with partial resistance in the absence of silicon; and downy mildew (*Bremia lactucae*) was significantly reduced in lettuce plants grown in a hydroponic culture amended with silicon (Garibaldi et al. 2012). Finally, the numbers of lesions caused by *Mycosphaerella pinodes* was reduced by 40–70 % in the leaves of pea plants amended with silicon (Dann and Muir 2002).

Strawberries The development of powdery mildew (*Sphaerotheca aphanis* var. *aphanis*) was reduced on plants grown in a hydroponic system containing silicon (Kanto et al. 2004). According to Kanto et al., disease severity for non-treated control plants had increased up to 17 times from its initial levels at 2 months after fungal inoculation compared with plants amended with silicon. Silicon concentrations were approximately 24 times greater in amended control plants, and disease severity decreased significantly when foliar silicon concentrations reached 1.5 dag/kg in that study (Kanto et al. 2004). A soil saturated with soluble silicon reduced powdery mildew development more effectively when used preventively as opposed to applying it after the initial incidence of disease, particularly for a susceptible cultivar over a highly susceptible one (Kanto et al. 2006). In that study, powdery mildew incidence was reduced by up to 86 % in plants from the highly susceptible cultivar and up to 58 % for plants from the less susceptible cultivar (Kanto et al. 2006). Pestalotia leaf spot (*Pestalotia longisetula*) severity was reduced by 61 % by the foliar application of 30 g/L of potassium silicate (Carré-Missio et al. 2010). According to these authors, the foliar application of potassium silicate was as effective as Acibenzolar-S-Methyl, a known inducer of host resistance, in reducing the symptoms of Pestalotia leaf spot, especially if applied foliarly before fungal inoculation.

Flowers Rose plants grown in soil amended with silicon showed a two- to four-fold increase in foliar silicon concentrations compared with plants grown in unamended soil (Shetty et al. 2012). The high foliar silicon concentration delayed powdery mildew (*Podosphaera pannosa*) onset by 1–2 days and reduced severity by up to 49 % (Shetty et al. 2012). Disease severity was also greatly reduced in the most resistant rose genotypes, which were measured to have the greatest levels of foliar silicon concentrations. However, powdery mildew (*Podosphaera fusca*) severity was not reduced in gerbera daisy plants grown in a substrate amended with silicon (Moyer et al. 2008).

Cotton Cotton plants grown in a hydroponic culture containing silicon and inoculated with *Phakopsora gossyppi* showed an increase in the incubation period and latent period, as well as reductions in the area under the rust progress curve, number of pustules and uredia per cm^2 of leaf area, and pustule area (Guerra et al. 2013a). Guerra et al. (2013b) also reported an increase in the incubation period of ramulosis (*Colletotrichum gossypii* var. *cephalosporioides*) and a decrease in the area under the disease progress curve for cotton plants grown in a hydroponic culture containing silicon. Cotton plants grown in a hydroponic culture containing silicon and inoculated with *Ramularia areola* also showed reductions of up to 35 % in the area under the ramularia leaf spot progress curve compared with non-amended plants (Curvêlo et al. 2013).

Grape The application of potassium silicate at a concentration of 1.7 mM to soil did not reduce the number of powdery mildew (*Uncinula necator*) colonies on grape leaves, while foliar sprays with the same potassium silicate concentration reduced the number of powdery mildew colonies by more than 60 % (Bowen et al. 1992). Blaich and Grundhöfer (1998) achieved a minor but significant increase in the resistance of six grape cultivars to powdery mildew when plants were grown in a nutrient solution containing 10 and 112 ppm of silicon dioxide. These authors concluded that silicon was necessary to increase grape resistance to *U. necator* infection; however, tissue susceptibility to the pathogen could not be eliminated solely by the exogenous application of silicon.

Bacterial Diseases

Bacterial Blight (*Xanthomonas oryzae* **pv.** *oryzae* (*Xoo*) **on Rice** The rice breeding lines TN1 and TSWY7, which are bacterial blight-susceptible and -resistant, respectively, developed shorter and smaller lesions when grown in a hydroponic culture containing silicon (Chang et al. 2002). A higher foliar soluble sugar content was positively correlated with disease development; however, under a higher silicon concentration, soluble foliar sugar levels were reduced considerably. This finding suggested that silicon created an internal environment unsuitable for bacterial growth. In a separate field experiment, three levels of calcium silicate (0, 2 and 4 t/ha) were applied to soil where four cultivars of rice with varying degrees of resistance to bacterial blight were grown (Chang et al. 2002). According to these authors, bacterial blight severity significantly decreased as calcium silicate concentrations increased. Feng et al. (2010) reported that bacterial blight severity in rice plants supplied with silicon decreased from 12 % to 52 % compared with non-amended plants.

Leaf Streak (*Xanthomonas translucens* pv. *undulosa*) on Wheat The wheat cultivar BR-18, which is susceptible to leaf streak, was grown in plastic pots containing a silicon-deficient soil amended with either calcium silicate or calcium carbonate (Silva et al. 2010). The foliar silicon concentration increased by 97 % for plants supplied with calcium silicate compared with those that received calcium carbonate. No difference in foliar calcium concentration was detected in plants that received the calcium silicate versus the calcium carbonate treatment; therefore, variations in silicon accounted for the observed differences in the level of resistance to leaf streak. The chlorotic leaf area was reduced by 50 % for plants supplied with calcium silicate compared with those receiving calcium carbonate. However, no differences were detected between plants that received the calcium silicate and calcium carbonate treatments for incubation period, latent period, necrotic leaf area or severity. Further, no differences in the bacterial leaf populations were noticed for either treatment; however, the values appeared to be lower at 4–8 days after inoculation for plants supplied with calcium silicate.

Bacterial Speck (*Pseudomonas syringae* pv. *tomato* (*Pst*)) on Tomato Tomatoes were grown in a soil without calcium silicate (control), in a soil without calcium silicate but sprayed with potassium silicate at 2 mL/L, and in a soil amended with calcium silicate (CS) at 0.16 g/kg and then inoculated with *Pst* (Andrade et al. 2013). The effect of potassium silicate on the growth of *Pst* was evaluated *in vitro*. No significant differences were observed among the treatments for foliar silicon concentration or the incubation period. In addition, no significant differences were detected for the number of lesions per plant or bacterial speck severity between the control and calcium silicate. However, the number of lesions per plant was significantly reduced by the foliar application of potassium silicate; and a negative linear response to increasing concentrations of potassium silicate was observed for *in vitro* growth of *Pst* (Andrade et al. 2013).

Bacterial Spot (*Xanthomonas axonopodis* pv. *passiflorae*) on Yellow Passion Fruit Brancaglione et al. (2009) evaluated the effect of silicate clay (0.5 %, 1.0 %, 1.5 % and 2.0 %) applied foliarly to yellow passion fruit seedlings as a preventive or curative control for bacterial spot. Plants were inoculated with the bacteria at 24 h before or after the application of the curative and preventive treatments, respectively. The silicate clay reduced the severity of bacterial spot in the preventive and curative treatments by approximately 100 % and 86 %, respectively, compared with the control plants.

Bacterial Fruit Blotch (*Acidovorax citrulli*) on Melon Conceição et al. (2014) evaluated the effect of silicon and antagonistic yeasts on the control of bacterial fruit blotch in melon seedlings and plants. In the first experiment, melon seedlings and plants were grown in a substrate amended with calcium silicate at a rate of 1.41 g of silicon/kg, and then were sprayed or not sprayed with yeasts. In the second experiment, seedlings and plants received a foliar treatment of yeast and potassium silicate

(1.7 mM silicon), either separately or in combination. Eighteen and fourty-three percent reductions were observed for the area under the disease progress curve of seedlings and plants grown in the silicon-amended substrate, respectively; and the area under the disease progress curve decreased by 81 % and 71 % for seedlings and plants sprayed with potassium silicate, respectively.

Viral Diseases

Tobacco *Nicotiana tabacum* plants were grown in a hydroponic culture containing silicon and then inoculated with *Tobacco ringspot virus* (TRSV) and *Tobacco mosaic virus* (TMV). Although *N. tabacum* is considered to be a low silicon accumulator species, the silicon amendment delayed the development of systemic TRSV symptoms and reduced the symptomatic leaf area in treated compared with untreated plants (Zellner et al. 2011). According to these authors, the silicon effect appeared to be virus-specific, because TMFV symptoms were not altered by the presence of silicon. In addition, the authors reported that foliar silicon concentrations for plants supplemented with the element increased up to fourfold when the plants were inoculated with TRSV compared with the control. This study provides the first evidence of silicon-primed plants dramatically increasing their accumulation of silicon when attacked by a virus. Furthermore, the influence of TRSV infection on foliar silicon concentrations suggests that the foliar accumulation of this element might be regulated in tobacco and may modulate the plant's defenses as part of a specific defense mechanism(s) against TRSV, especially because silicon apparently does not act directly on the infectivity of TRSV particles.

Interaction of Silicon and Fungicides

Soil applied calcium silicate slag reduced rice panicle blast severity almost as effectively as a mercuric fungicide (12 vs. 10 % and 11.2 vs. 7.4 %, respectively), depending on the level of nitrogen (N) applied (Kitani et al. 1960). According to these authors, calcium silicate was associated with a gain in grain weight over the control by 37 % (50 kg N/ha) to 40 % (75 kg N/ha). The mercuric fungicide increased weights by 28 % and 34 % for the two N treatments, respectively. Combined silicon and fungicide treatments were most effective in reducing panicle blast severity (below 3 %) and increasing grain weight (40–48 %). Hashimoto and Hirano (1976) conducted similar studies on panicle blast development but included other factors such as rice cultivars and nitrogen. According to their data, calcium silicate alone reduced rice blast by 13 %, fungicide alone (Hinosan) reduced blast by 22 % and the combined fungicide + calcium silicate treatment reduced blast by 27 % in comparison with the non-amended control. In Florida, an evaluation of

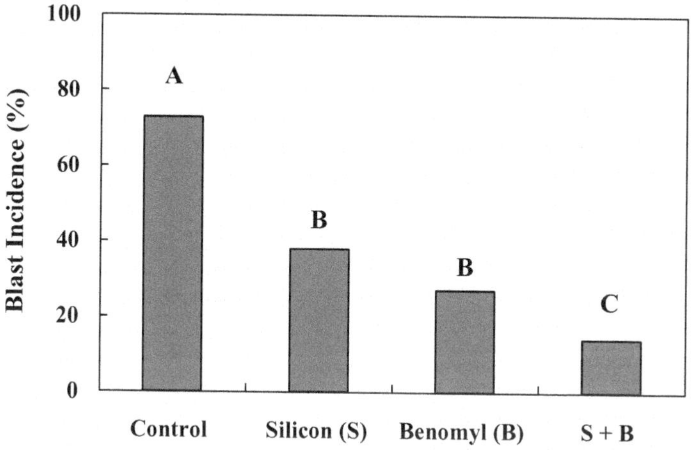

Fig. 4.7 Influence of silicon and benomyl on panicle blast incidence. Values with the same letter are not significantly different based on Fisher's LSD ($P=0.05$) (Reproduced from Datnoff et al. (1997), with permission from Elsevier)

silicon fertilization in combination with benomyl or propiconazole was undertaken to determine whether silicon could control diseases such as blast or brown spot as effectively as a fungicide (Datnoff and Snyder 1994; Datnoff et al. 1997). A rice crop was treated with silicon at 0 and 2 t of silicon/ha, benomyl at 0 and 1.68 kg/ha, and propiconazole at 0 and 0.44 L/ha. Fungicides were applied at panicle differentiation, boot, heading and heading plus 14 days. Blast incidence was 73 % for plots not treated with silicon or fungicide (control) and 27 % in the benomyl treated plots (Fig. 4.7). Where silicon was applied, blast incidence was 36 % in the non-fungicide plots and 13 % in the benomyl-treated plots (Fig. 4.7). The same degree of blast control was generally obtained when either benomyl or silicon were applied individually. Brown spot responses were similar to those observed for blast (Fig. 4.8). Brown spot severity and disease progress were reduced more by silicon alone than with propiconazole. For both diseases, the greatest reduction in disease development was obtained by integrating silicon fertilization with fungicides.

Mathai et al. (1977) evaluated the effect of silicon applied as sodium silicate alone and in combination with two fungicides, Hinosan and Dithane 45, on the control of sheath blight. All treatments were effective in reducing sheath blight intensity (SBI) and increasing yields in comparison with the control; silicon (SBI=48 % and yield=4.6 %), Dithane (SBI=68 % and yield=9.5 %), Hinosan (SBI=99 % and yield=16.8 %), Dithane+silicon (SBI=84 % and yield=13.1 %) and Hinosan+silicon (SBI=118 % and yield=37.2 %). The combination of silicon and a fungicide was the best treatment to reduce SBI. The increase in grain yield was synergistic when the fungicide Hinosan was used in combination with silicon.

Because silicon can control several rice diseases to the same general degree as a fungicide, this element could help decrease the number of fungicide applications or the concentration of the active ingredient. This hypothesis was tested by Seebold

Fig. 4.8 Symptoms of brown spot as influenced by applications of silicon (**a** and **e**), propiconazole (**b**), the combination of silicon + propiconazole (**c**) and the nontreated control (**d**) (Reproduced from Datnoff et al. (1997), with permission from Elsevier)

et al. (2004) under field experiments with upland rice in the savannahs of Colombia. Silicon was applied as wollastonite at 400 kg elemental silicon/ha. Treatments included an untreated control, silicon applied alone and silicon plus fungicides (edifenphos at 1 L/ha and tricyclazole at 300 g/ha) applied during the following growth stages: tillering (T), panicle initiation (PI), booting (B), 1 % panicle emergence (1 %), 50 % panicle emergence (50 %); PI, B, 1 %, and 50 %; B, 1 % and 50 %; 1 % and 50 %; B and 1 %; PI and 1 %; and T. The incidence of panicle blast was significantly reduced using both silicon alone and silicon plus fungicides in comparison with the untreated control (Fig. 4.9). Silicon alone significantly reduced the incidence of panicle blast by 40 %. The silicon plus one fungicide treatment reduced panicle blast by 75–90 %, while the application of silicon plus two fungicide treatments reduced panicle blast from 76 % to 94 %. Silicon plus three to five fungicide treatments reduced panicle blast 94–98 % (Fig. 4.9). Therefore, one application of the fungicide in combination with silicon was as effective as two, with increased results observed for 3–5 applications. No significant differences in yield were observed between the silicon alone or silicon plus fungicide applications, regardless of timing, and all treatments significantly increased yield in comparison with the

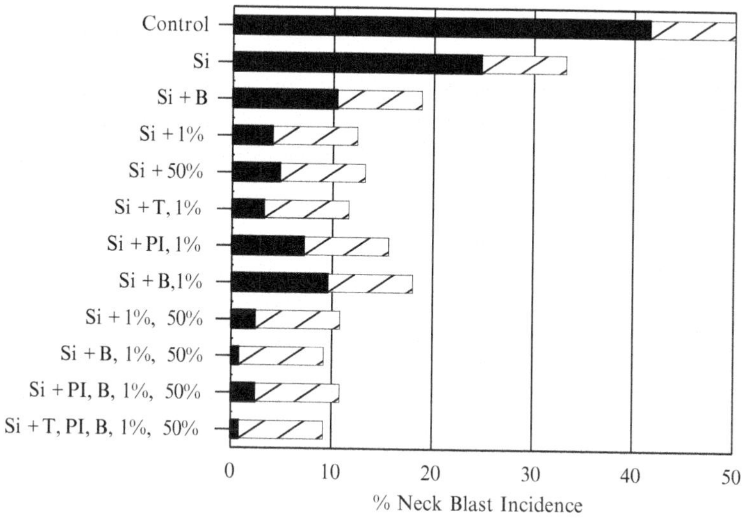

Fig. 4.9 Effect of silicon and fungicide timings on panicle (neck) blast incidence. Fungicides timings are: tillering (*T*), panicle initiation (*PI*), booting (*B*), 1 % heading (1 %), 50 % heading (50 %) and various combinations. Stripe bars represent FLSD value ($P = 0.05$) (Reproduced from Datnoff and Rodrigues (2005))

control. In another experiment, silicon was incorporated prior to seeding at 0 and 1000 kg elemental silicon/ha (Seebold et al. 2004). Two foliar applications of edifenphos were applied at 0 %, 10 %, 25 % and 100 % of the manufacturer's recommended concentration. Leaf blast levels for silicon alone and silicon plus edifenphos at various concentrations were 54–75 % lower than in the untreated control. For panicle blast, silicon alone and silicon plus edifenphos and tricyclazole at various concentrations resulted in incidence levels 28–66 % lower than the untreated control. The greatest leaf and panicle blast reductions were observed where silicon plus the full concentration of fungicide had been applied. Silicon plus lower rates of the fungicides (10 % and 25 %) were able to reduce leaf and panicle blast as effectively as a full rate of each fungicide. Silicon alone was just as effective as the fungicides alone in reducing leaf blast severity and promoting plant growth when compared with the control treatment (Fig. 4.10). Fungicides improved yields 22–28 % over the control. Interestingly, silicon alone improved yields by 51 %, which was significantly greater than the fungicides' contribution. The effect of silicon on reducing a disease such as blast unquestionably contributes to an increase in yield, but silicon also has been shown to increase yields in the absence of disease (Ou 1985).

In 1995 and 1996, silicon was added to soil at 0 and 1000 kg elemental silicon/ha prior to seeding (Seebold et al. 2004). Plots that were treated in 1995 (residual silicon) were compared to plots receiving a fresh or current year treatment of silicon in 1996 to study the residual effect. Two foliar applications of edifenphos, sprayed at 20 and 35 days after planting, were performed followed by three applications of tricyclazole. Leaf blast was evaluated as percent area of individual leaves, and

Fig. 4.10 Overall symptoms of blast on the lower leaves of the rice cultivar Oryzica 1 for the non-amended control (**a**) in comparison to plants amended with silicon (**b**) or treated with fungicides (**c**). Note the clear difference in plant vigor between the non-amended control and the treatments receiving either silicon or fungicides (Reproduced from Datnoff and Rodrigues (2005))

panicle blast was rated as percent incidence of 100 panicles. In both 1995 and 1996, leaf blast levels in plants treated with silicon alone (residual and fresh applications) and silicon (residual and fresh applications) plus edifenphos were 50–68 % lower in comparison with the untreated control (Seebold et al. 2004). The greatest reductions in leaf blast were observed where silicon plus fungicide had been applied. The 1 year residual silicon application was as effective as a fresh application, and these treatments were not significantly different for leaf blast control in comparison to edifenphos alone or in combination with a 1 year residual. Silicon alone reduced leaf blast to the same level as edifenphos applied with silicon in 1995. In 1996, leaf blast levels for plants treated with silicon alone were significantly lower (35 %) than for those treated with the full rate of fungicide. The incidence of panicle blast was reduced by 28 % and 66 % with applications of silicon and silicon plus tricyclazole. A 1 year residual application of silicon applied in 1995 was as effective as a fresh application in 1996 in reducing panicle blast incidence. However, these treatments were not as effective as a fungicide applied alone or in combination with silicon.

The fungicide tricyclazole alone or in combination with silicon provided the best reductions in panicle blast incidence. Silicon alone and in combination with tricyclazole applied in 1995 or in 1996 increased yields 28–51 % over the untreated control. The 1995 residual silicon application was as effective at increasing yields as the 1996 application and did not differ significantly from the tricyclazole alone or tricyclazole applied in combination with the 1995 or 1996 silicon application (Seebold et al. 2004).

Resende et al. (2013) evaluated the effect of silicon and its interaction with a fungicide on the management of anthracnose on sorghum. The experiments were conducted in a silicon-deficient soil during the 2008/2009 and 2009/2010 growing seasons. Calcium silicate and lime applied at the concentrations of 6 and 5 t/ha, respectively, were randomly assigned to the main plot. Two sorghum lines, BR-008 (resistant) and BR-009 (susceptible), were assigned to the split plots. The split-split plots corresponded to the presence or absence of the fungicide combination epoxiconazole + pyraclostrobin. The residual effect of calcium silicate and lime from the 2008/2009 growing season was evaluated in the 2009/2010 growing season. For the 2008/2009 growing season, the area under the anthracnose progress curve (AUAPC) was reduced by 39 % and 42 % for lines BR-008 and BR-009, respectively, in the presence of calcium silicate. In the presence of the fungicide, the AUAPC was reduced by 35 % and 42 % for the calcium silicate and lime treatments, respectively. Calcium silicate in the presence and absence of the fungicides contributed to a decrease in the AUAPC by 44 % and 37 %, respectively. The fungicide application alone decreased the AUAPC by 50 % and 39 % for lines BR-008 and BR-009, respectively. Without the fungicide, the AUAPC decreased by 88 % for line BR-008 compared with line BR-009; with a fungicide, the reduction reached 90 %. Foliar silicon concentrations significantly increased with the calcium silicate application (5.9 g/kg) compared with the lime application (0.3 g/kg), regardless of the sorghum line. Yield increased by 0.6 t/ha for the calcium silicate compared with the lime-treated crop. The fungicide increased yield by 0.48 t/ha compared with the non-fungicide treatment. The residual effect of calcium silicate in the soil increased foliar silicon concentrations and yield, as well as reduced the intensity of anthracnose in the 2009/2010 growing season. The results of this study combined with previous reports from other pathosystems support the conclusion that anthracnose intensity can be reduced and yield can be increased when sorghum plants, especially from a susceptible line, are grown in a silicon-deficient soil amended with calcium silicate.

Polanco et al. (2014) evaluated the effect of foliar applications of potassium silicate (KSi), both alone and in combination with the fungicide (F) azoxystrobin, in the control of anthracnose and, subsequently, on bean yield. KSi was applied at 20, 27, 40 and 55 days after sowing (das) and the fungicide was applied at 27, 40 and 55 days. The area under the disease progress curve was reduced by 21 %, 64 % and 76 % for KSi, F and KSi plus F sprays, respectively, while mean yield increased by 10 %, 138 % and 148 %, respectively, in experiment 1. For experiment 2, AUDPC was reduced by 36 %, 62 %, and 74 % for KSi, F and KSi plus F treatments, respectively, while mean yield increased by 13 %, 102 % and 153 %, respectively. The results of this study suggest the possibility of using a foliar application of KSi in association with a fungicide to reduce anthracnose severity on bean plants and, subsequently, achieve greater gains in yield due to improved plant growth. Rodrigues et al. (2015) investigated whether foliar sprays of KSi, sodium molybdate (NaMo) or a combination of both (KSi plus NaMo), with or without the fungicide azoxystrobin (Azox), could reduce anthracnose symptoms, improve photosynthesis and increase yield. The treatments were as follows: (i) KSi; (ii) NaMo; (iii) KSi plus NaMo; (iv) Azox; (v) Azox plus KSi; (vi) Azox + NaMo; (vii) Azox plus KSi + NaMo, and (viii) control (no KSi, NaMo or Azox). The KSi, NaMo and Azox treatments were applied at concentrations of 35 g/L, 90 g/ha and 120 g ai/ha, respectively. The KSi was applied at 20, 27, 40 and 55 das. The NaMo was applied only at 27 das, and the fungicide was applied at 27, 40 and 55 das. The plants were inoculated with *C. lindemuthianum* at 23 das. Anthracnose severity was reduced by 64 % and yield increased by 156 % in plants sprayed with the fungicide compared with those not sprayed. The KSi, NaMo and NaMo plus KSi applications reduced anthracnose severity by 32 %, 16 % and 38 %, respectively, while yield increased by 17 %, 19 % and 64 %, respectively.

Duarte et al. (2008) investigated the effect of the foliar application of potassium silicate, alone or mixed with different rates of fungicides, on the control of potato late blight. The fungicide cimoxanil + mancozeb (60 + 700 g/Kg a.i.) and 60 g/L of potassium silicate (pH 5.5) were used. The treatments (T) were: T1 – Control; T2 – cimoxanil + mancozeb (2.0 Kg/ha); T3 – cimoxanil + mancozeb (2.5 Kg/ha); T4 – cimoxanil + mancozeb (2.0 Kg/ha) + potassium silicate; T5 – cimoxanil + mancozeb (2.5 Kg/ha) plus potassium silicate; T6 – cimoxanil + mancozeb (3.0 Kg/ha) plus potassium silicate; T7 – potassium silicate and T8 – cimoxanil + mancozeb (3.0 Kg/ha). Treatments were applied weekly. The areas under the disease progress curves were 73, 24, 18, 30, 20, 18, 68 and 16 for the treatments T1, T2, T3, T4, T5, T6, T7, and T8, respectively. The potassium silicate treatment was not effective in decreasing late blight development and did not show any additional benefit when mixed with a fungicide.

Wordell Filho et al. (2013) evaluated the effects of the foliar application of potassium silicate and fungicides (azoxystrobin 60 g a.i/ha + ciproconazole + 24 g a.i/ha) for the control of leaf rust (*Puccinia triticina*) and yellow spot (*Drechslera tritici-repentis*) on the wheat cultivars Safira and Quartzo. The cultivars were randomly assigned to the main plot and the treatments T1 – control; T2 – fungicide sprayed at growth stages 45 (booting) and 58 (heading); T3 – potassium silicate (40 g/L) sprayed at growth stages 45 and 58; T4 – potassium silicate (40 g/L) and fungicides sprayed at plant growth stages 45 and 58, respectively; and T5 – potassium hydroxide (6.5 g/L) sprayed at growth stages 45 and 58 were applied to the subplots. Potassium silicate, regardless of the concentration, number of foliar applications or wheat cultivar, was not effective in decreasing either the area under the leaf rust progress curve or the area under the yellow spot progress curve.

Lopes et al. (2014b) evaluated the effects of various calcium silicate concentrations combined with the fungicide triadimenol on the incidence of coffee leaf rust. Calcium silicate (CS) and lime (L) were used according to the following mixture combinations (M): M1: 0 % CS and 100 % L; M2: 25 % CS and 75 % L; M3: 50 % CS and 50 % L; M4: 75 % CS and 25 % L; and M5: 100 % CS and 0 % L. Foliar silicon concentrations did not increase as CS concentrations increased in the soil. There was no reduction in the area under the rust progress curve as the concentrations of CS increased in the soil. During the 2006/2007, 2007/2008 and 2008/2009 growing seasons, rust incidence reached 94 %, 96 % and 92 % on plants that did not receive triadimenol, respectively; whereas the incidence did not exceed 6 %, 38 % and 16 %, respectively, for those plants that received the fungicide. The 3-year experiment indicated that soil amendments using calcium silicate had no effect on reducing coffee leaf rust incidence, likely because the roots of this plant lack transporter genes to move the element to the leaves, thereby increasing the foliar silicon concentration. Conversely, coffee leaf rust symptoms were dramatically reduced in plants treated with triadimenol.

In conclusion, silicon can control diseases as effectively as fungicides in some pathosystems and can often help to reduce the number of fungicides applied in a growing season. The use of silicon plus reduced rates of fungicides are as effective as full rates of fungicides alone. These results suggest that the number of fungicide applications and their rates may be reduced. A 1 year residual silicon application was also effective in reducing blast in rice and anthracnose in sorghum and maintained yields in both crops. Silicon alone enhanced rice yields more effectively than fungicides alone. Consequently, growers may save either initial or additional application costs for either fungicides or silicon while providing positive environmental benefits (Alvarez and Datnoff 2001).

The Efficiency in Applying Silicon via the Foliage versus the Root in Controlling Plant Disease

The use of silicon has gained much attention from the scientific community because of its potential to control plant diseases in a more sustainable and environmental friendly way. This element can be applied to plants either through soil amendments for solid sources or via foliar sprays for liquid sources. Liquid foliar products containing readily available soluble silicon are not taken up by foliage because transporter genes have not been reported to exist in this plant organ. Consequently, liquid sources quickly polymerize on the leaf surface and form a physical barrier that may affect fungal penetration and sporulation to lesser degrees than root uptake would (Liang et al. 2005; Dallagnol et al. 2012; Cacique et al. 2013). Soil-based silicon amendments, particularly for root applications, can be applied as calcium silicate or other solid/liquid silicon sources through their incorporation into the soil prior to sowing or as soluble silicon (e.g. silicate salts) mixed into the nutrient solution when plants are grown in hydroponic systems. Soluble silicon sources can also be diluted in water and applied to the shoots as foliar sprays. The efficiency of foliar and root applications in suppressing disease vary depending on the plant species, the environment where plants are cultivated, as well as host-parasite interactions.

When silicon is supplied to plants through the roots, this element will move through the xylem into the root endodermis, cell membranes of the vascular bundle and leaf cells of the epidermis just beneath the cuticle. In contrast, when silicon is applied foliarly, deposition occurs on leaf surfaces and may be easily removed by rain or irrigation water (Dallagnol et al. 2012; Guével et al. 2007; Kim et al. 2002; Rezende et al. 2009). The mechanism of action used by the plants to defend themselves against infection by certain pathogens greatly depends on the mode of silicon application (root versus foliar). A more detailed discussion about how silicon can reduce the intensities of diseases is provided in the Chap. 5. Briefly, when present inside plant tissue, silicon may act both by forming a physical barrier and by potentiating biochemical mechanisms involved in host defense. However, silicon applied foliarly may act by forming a chemical-physical barrier (i.e., by changing the pH or the osmotic potential) on the leaf surface.

Despite the differences in silicon distribution either in or on plant tissue based on how the element is applied, both root and foliar silicon applications will decrease disease development. In the melon-*Podosphaera xanthii* interaction, both root and foliar applications of silicon affected all epidemic components when compared with the unamended control, except for the latent period. However, the effects of silicon applied to the roots were more pronounced in comparison with those observed when silicon was applied foliarly, resulting in greater reductions in the rate of colony expansion, colony area and conidial production (Dallagnol et al. 2012). According to these authors, the effects of both forms of silicon application on the components of host resistance correlated with lower values for the area under the disease progress curve; but the effect on this variable was greater when silicon was supplied to the roots. For melon plants exposed to natural infection and evaluated over 60 days, the application of silicon reduced the disease progress rate in the lower, middle, and upper plant parts, irrespective of the form of application. However, the root application was significantly more effective than the foliar treatment in reducing the disease progress rate (Dallagnol et al. 2012). Liang et al. (2005) used two cucumber cultivars that differed in their resistance to powdery mildew to study the effects of applying silicon foliarly and to the roots against infection by *P. xanthii*. In both foliar and root applications of silicon, plants were subjected to one or two pathogen inoculations. The authors observed that the root silicon amendment significantly reduced disease severity regardless whether the plants received one or two inoculations. In addition, applying silicon to the roots increased the efficacy of disease suppression when the plant was subjected to a second pathogen inoculation. In contrast, the foliar application was ineffective in reducing disease severity regardless whether the plants received one or two pathogen inoculations. These results demonstrate that the greatest disease control can be achieved by a continuous supply of silicon via the roots, which subsequently enhanced the plant's defense system. Although the foliar application of silicon was effective in reducing infection by *P. xanthii*, the mechanism was most likely the creation of a physical barrier through the deposition of material on the leaf surface. This deposition likely caused a change in surface pH and/or osmotically, but did not enhance the plant's defense responses. Consequently, protection from fungal infection would only take place where the plant tissue received the applied silicon directly (Liang et al. 2005). A similar effect was also reported for melon plants inoculated with *P. xanthii* (Dallagnol et al. 2012). In this study, the leaf surface areas that did not receive applications of potassium silicate had fungal colony development rates similar to those observed for the untreated controls. This observation supported earlier findings that disease control was associated only with the deposition of potassium silicate on the leaf surface (Dallagnol et al. 2012). In zucchini, silicon applied foliarly reduced the severity of powdery mildew (*P. xanthii*); however, the efficacy of silicon was improved by increasing the application frequency from one to three times per week (Tesfagiorgis and Laing 2011). Furthermore, when runoff of the silicon solution from the treated leaves reached the root zone of the plants, better disease control was achieved. This suggested that silicon was taken up by the roots, which activated host defense responses against fungal infection (Tesfagiorgis and Laing 2011). Wolff et al. (2012) also reported an increase in the inhibition of *P. xanthii* growth on cucumber leaves by increasing the frequency of foliar applications of silicon. However, the authors noted that careful attention should be given to high dose applications of foliar sprays from different silicon sources because they may reduce leaf gas exchange and yield, despite being effective in controlling powdery mildew (Ramos et al. 2013).

Rodrigues et al. (2009) tested the efficacy of potassium silicate foliar sprays in controlling soybean rust and showed, for the first time, that disease severity in this plant could be reduced under both greenhouse and field conditions. Lemes et al. (2011) compared both root and foliar silicon source applications in the control of soybean rust. Experimental greenhouse results demonstrated that both silicon application methods delayed disease onset by approximately 3 days and decreased the area under the disease progress curve. However, in field experiments an average delay in disease onset of 3 days was only observed for root silicon treatments, and root treatments were also more effective in reducing the area under the disease progress curve. The authors of the study surmised that the absence of any significant differences between root and foliar silicon application on the area under the disease progress curve under greenhouse conditions was likely due to the low disease pressure observed under greenhouse conditions compared with what occurred in the field. Under field conditions, silicon soil amendments were more effective than silicon applied foliarly in suppressing soybean rust development because the soil treatments resulted in both a delay in disease onset as well as a reduction in the area under the disease progress curve. Interestingly, foliar applications of silicon did not result in an increase in foliar silicon concentrations (Lemes et al. 2011). However, the efficacy of soybean rust control by silicon was recently shown to be dependent on the ability of soybean plants to accumulate this element, and an innate variability in the ability to uptake silicon existed within soybean germplasm (Arsenault-Labrecque et al. 2012). According to Arsenault-Labrecque et al. (2012), soybean cultivars that showed no significant differences in foliar silicon concentrations, regardless of the type of silicon amendment, also did not show any differences in soybean rust incidence. However, a near absence in symptoms of soybean rust was observed for the Korean cultivar Hikmok sorip when supplied with silicon, and this cultivar accumulated nearly four times more silicon than 'Williams 82'. Although Cruz et al. (2013) found an increase in silicon concentrations in soybean leaves when plants were amended with silicon via roots, no significant differences in disease control were detected between root and foliar silicon applications. Moreover, on plants sprayed with silicon, the uredia were smaller and more compact than those observed on the leaves of plants amended with silicon through the roots. In another study, Cruz et al. (2014b) showed that foliarly-applied silicon was more effective than applying silicon to the roots in reducing soybean rust symptoms, even when applying silicon to the roots increased the incubation period of the disease.

The efficacy of both leaf and root silicon applications were compared for blast and brown spot control in rice. For brown spot, the foliar application of silicon decreased disease intensity; however, the level of control achieved was not as great as that obtained when silicon was supplied to the roots (Rezende et al. 2009). For blast, lesion size, the number of lesions per cm^2 of leaf area and the area under the blast progress curve were reduced for both methods of silicon application, but silicon supplied to the roots tended to be more effective in suppressing blast development than that applied foliarly (Cacique et al. 2013). Abed-Ashtiani et al. (2012) also reported that both methods of silicon application significantly reduced blast severity, but comparing both methods indicated that the root application resulted in greater rates of disease control.

Foliar silicon applications to wheat were not as effective as applying silicon to the roots in controlling powdery mildew (Guével et al. 2007). According to Guével et al., X-ray microanalyses of treated plants revealed that root applications resulted in a more consistent leaf silicon deposition. Furthermore, root applications consistently yielded the best disease severity reduction rates, leading to a decrease as high as 80 % while leaf application only led to a reduction of approximately 40 %.

Foliar silicon applications may still be an attractive alternative to control diseases in an environmental friendly way for crops such as tomato and coffee, which do not actively accumulate silicon in their shoots. For tomato, silicon applied foliarly was ineffective in controlling late blight (Duarte et al. 2007), but it was highly effective in reducing powdery mildew severity up to 90 % under field conditions (Yanar et al. 2011). For coffee, potassium silicate applied foliarly reduced rust severity (Pereira et al. 2009; Carré-Missio et al. 2012b) and the severity of Cercospora leaf spot up to 30 % under both field and greenhouse conditions (Amaral et al. 2008). Carré-Missio et al. (2014) observed that plates of polymerized potassium silicate on the leaf surface of coffee plants contributed to the reduction in fungal colonization and the number of uredia produced by *Hemileia vastatrix*. The authors concluded that the foliar application of silicon controlled coffee leaf rust development through the creation of a physical barrier in the form of polymerized potassium silicate, its osmotic effect against urediniospore germination or both.

In summary, silicon applied via the roots is consistently more effective than foliarly-applied silicon in the control of plant diseases, but this mode of application depends on a plant's ability to uptake silicon from the soil solution and accumulate it in the shoot. However, foliar application of silicon can still be used for many plant species, with the understanding that the efficacy for disease control may depend on the application frequency, environmental conditions, the life cycle of the pathogen and its level of aggressiveness, as well as the amount of inoculum produced during the pathogen's life cycle. The efficacy of foliar and root applications of silicon in disease control is summarized in Table 4.2.

Table 4.2 Comparative efficiency of foliar and root applications of soluble silicon in controlling certain foliar diseases

Crops	Diseases	Pathogens	Efficiency in disease control		Variables evaluated	Environment	Inoculation	References
			Root	Leaf				
Cucumber	Powdery mildew	*Podosphaera xanthii*	15–36 %	Ineffective	Severity	Greenhouse	Artificial	Liang et al. (2005)
Melon	Powdery mildew	*Podosphaera xanthii*	73 %	65 %	AUDPC	Greenhouse	Artificial	Dallagnol et al. (2012)
Rice	Blast	*Pyricularia oryzae*	80–95 %	52–91 %	AUDPC	Greenhouse	Artificial	Cacique et al. (2013)
Rice	Blast	*Pyricularia oryzae*	75 %	53 %	Incidence	Greenhouse	Artificial	Abed-Ashtiani et al. (2012)
Rice	Blast	*Pyricularia oryzae*	76 %	56 %	Severity	Greenhouse	Artificial	Abed-Ashtiani et al. (2012)
Rice	Brown spot	*Bipolaris oryzae*	37 %	Ineffective	AUDPC	Greenhouse	Artificial	Rezende et al. (2009)
Soybean	Rust	*Phakopsora pachyrhizi*	5–54 %	33–89 %	Severity	Greenhouse	Artificial	Cruz et al. (2013, 2014), Lemes et al. (2011)
Soybean	Rust	*Phakopsora pachyrhizi*	16–43 %	19–36 %	Severity	Field	Natural	Lemes et al. (2011)
Wheat	Powdery mildew	*Blumeria graminis* f. sp. *tritici*	~80 %	~40 %	Severity	Greenhouse	Artificial	Guével et al. (2007)

AUDPC = area under disease progress curve

Conclusions

That silicon plays an important role in the mineral nutrition of many plant species is not in doubt, nor is its ability to efficiently reduce the intensities of several diseases. Effective, practical means of application and affordable sources of silicon are needed for use in agriculture, particular for row crops. As the need for environmentally friendly strategies for plant disease management increases, silicon nutrition could provide a valuable tool for use in crops able to uptake and accumulate silicon efficiently, such as rice. The use of silicon in the control of plant diseases would be well-suited for inclusion in an integrated disease management strategy and would permit possible reductions in the use of fungicides while enhancing host plant resistance. As researchers and growers become more aware of the importance of silicon in sustainable agriculture, it is likely that this often overlooked, quasi-essential element will be recognized as a viable means of managing important plant diseases.

References

Abed-Ashtiani F, Kadir JB, Selamat AB, Hanif AHBM, Nasehi A (2012) Effect of foliar and root application of silicon against rice blast fungus in MR219 rice variety. Plant Pathol J 28:164–171. doi:10.5423/PPJ.2012.28.2.164

Adatia MH, Besford RT (1986) The effects of silicon on cucumber plants grown in recirculating nutrient solution. Ann Bot 58:343–351

Aleshin NE, Avakyan ER, Dyakunchak SA, Aleshkin EP, Baryshok VP, Voronkov MG (1987) Role of silicon in resistance of rice to blast. Dokl Akad Nauk 291:217–219

Alvarez J, Datnoff LE (2001) The economic potential of silicon for integrated management and sustainable rice production. Crop Prot 20:43–48

Amaral DR, Resende MLV, Ribeiro PM Jr, Borel JC, Mac Leod REO, Pádua MA (2008) Silicato de potássio na proteção do cafeeiro contra *Cercopsora coffeicola*. Trop Plant Pathol 33:425–431. doi:10.1590/s1982-56762008000600004

Andrade CC, Resende RS, Rodrigues FA, Ferraz HGM, Moreira W, Oliveira JR, Mariano RLR (2013) Silicon reduces bacterial speck development on tomato leaves. Trop Plant Pathol 38:436–442. doi:10.1590/S1982-56762013005000021

Aquino LA, Berger PG, Rodrigues FA, Zambolim L, Ogoshi F, Miranda LM, Lélis MM (2008) Controle alternativo da mancha de Ramularia do algodoeiro. Summa Phytopathol 34:131–136. doi:10.1590/S0100-54052008000200004

Arsenault-Labrecque G, Menzies JG, Bélanger RR (2012) Effect of silicon absorption on soybean resistance to *Phakopsora pachyrhizi* in different cultivars. Plant Dis 96:37–42. doi:10.1094/PDIS-05-11-0376

Bélanger RR, Benhamou N, Menzies JG (2003) Cytological evidence of an active role of silicon in wheat resistance to powdery mildew (*Blumeria graminis* f. sp. *tritici*). Phytopathology 93:402–412. doi:10.1094/PHYTO.2003.93.4.402

Bengsch E, Korte F, Polster J, Schwenk M, Zinkernagel V (1989) Reduction in symptom expression of *Belladonna mottle* virus infection on tobacco plants by boron supply and the antagonistic action of silicon. Z Naturforsch 44:777–780

Berni RF, Prabhu NA (2003) Eficiência relativa de fontes de silício no controle de brusone nas folhas em arroz. Pesq Agrop Brasileira 38:195–201

Bi Y, Tian SP, Guo YR, Ge YH, Qin GZ (2006) Sodium silicate reduces postharvest decay on Hami melons: induced resistance and fungistatic effects. Plant Dis 90:279–283

Blaich R, Grundhöfer H (1998) The influence of silica fertilization on the resistance of grapevines to powdery mildew. Vitis 37:21–26

Botelho DMS, Pozza EA, Pozza AAA, Carvalho JG, Botelho CE, Souza PE (2005) Intensidade da cercosporiose em mudas de cafeeiro em função de fontes e doses de silício. Fitopatol Bras 30:582–588. doi:10.1590/S0100-41582005000600003

Bowen P, Menzies J, Ehret D, Samuels L, Glass ADM (1992) Soluble silicon sprays inhibit powdery mildew development on grape leaves. J Am Soc Hortic Sci 117:906–912

Brancaglione P, Sampaio AC, Fischer IH, Almeida AM, Fumis TF (2009) Eficiência de argila silicatada no controle de *Xanthomonas axonopodis* pv. *passiflorae*, *in vitro* e em mudas de maracujazeiro-amarelo. Rev Bras Frutic 31:718–724. doi:10.1590/S0100-29452009000300014

Brecht MO, Datnoff LE, Kucharek TA, Nagata RT (2007) The influence of silicon on the components of resistance to gray leaf spot in St. Augustinegrass. J Plant Nutr 30:1005–1021. doi:10.1080/01904160701394287

Cacique IS, Domiciano GP, Rodrigues FA, Vale FXR (2012) Silicon and manganese on rice resistance to blast. Bragantia 71:239–244. doi:10.1590/S0006-87052012000200013

Cacique IS, Domiciano GP, Moreira WR, Rodrigues FA, Cruz MFA, Serra NS, Català AB (2013) Effect of root and leaf applications of soluble silicon on blast development in rice. Bragantia 72:304–309. doi:10.1590/brag.2013.032

Camargo MS, Amorim L, Júnior ARG (2013) Silicon fertilization decreases brown rust incidence in sugarcane. Crop Prot 53:72–79. doi:10.1016/j.cropro.2013.06.006

Carré-Missio V, Rodrigues FA, Schurt DA, Pereira SC, Oliveira MGA, Zambolim L (2009) Ineficiência do silício no controle da ferrugem do cafeeiro em solução nutritiva. Trop Plant Pathol 34:416–421. doi:10.1590/S1982-56762009000600009

Carré-Missio V, Rodrigues FA, Schurt DA, Rezende DC, Ribeiro NB, Zambolim L (2010) Aplicação foliar de silicato de potássio, acibenzolar-S-metil e fungicidas na redução da mancha de *Pestalotia* em morango. Trop Plant Pathol 35:182–185. doi:10.1590/S1982-56762010000300008

Carré-Missio V, Rodrigues FA, Schurt DA, Moreira WR, Rezende DC, Korndörfer GH, Zambolim L (2012a) Proteção local, não sistêmica, do silicato de potássio reduz os sintomas da ferrugem do cafeeiro. Trop Plant Pathol 37:275–280. doi:10.1590/S1982-56762012000400007

Carré-Missio V, Rodrigues FA, Schurt DA, Rezende DC, Moreira WR, Korndörfer GH, Zambolim L (2012b) Componentes epidemiológicos da ferrugem do cafeeiro afetados pela aplicação foliar de silicato de potássio. Trop Plant Pathol 37:50–56. doi:10.1590/S1982-56762012000100006

Carré-Missio V, Rodrigues FA, Schurt DA, Resende RS, Souza NFA, Rezende DC, Moreira WR, Zambolim L (2014) Effect of foliar-applied potassium silicate on coffee leaf infection by *Hemileia vastatrix*. Ann Appl Biol 164:396–403. doi:10.1111/aab.12109

Chang SJ, Tzeng DDS, Li CC (2002) Effect of silicon nutrient on bacterial blight resistance of rice (*Oryza sativa* L.). In: Matoh T (ed) Second silicon in agriculture conference. Press-Net, Kyoto, pp 31–33

Conceição CS, Felix KCS, Mariano RLR, Medeiros EV, Souza EB (2014) Combined effect of yeast and silicon on the control of bacterial fruit blotch in melon. Sci Hortic 174:164–170. doi:10.1016/j.scienta.2014.05.027

Côté-Beaulieua C, Chaina F, Menzies JG, Kinrade SD, Bélanger RR (2009) Absorption of aqueous inorganic and organic silicon compounds by wheat and their effect on growth and powdery mildew control. Environ Exp Bot 65:155–161. doi:10.1016/j.envexpbot.2008.09.003

Cruz MFA, Silva LF, Rodrigues FA, Araujo JM, Barros EG (2012) Silício no processo infeccioso de *Phakopsora pachyrhizi* em folíolos de plantas de soja. Pesq Agrop Brasileira 47:142–145. doi:10.1590/S0100-204X2012000100020

Cruz MFA, Rodrigues FA, Diniz APC, Moreira MA, Barros EG (2013) Potassium silicate and calcium silicate on the resistance of soybean to *Phakopsora pachyrhizi* infection. Bragantia 72:373–377. doi:10.1590/brag.2013.052

Cruz MFA, Araujo L, Polanco LR, Rodrigues FA (2014a) Aspectos microscópicos da interação feijoeiro-*Colletotrichum lindemuthianum* mediados pelo silício. Bragantia 73:284–291. doi:10.1590/1678-4499.0139

Cruz MFA, Rodrigues FA, Diniz APC, Moreira MA, Barros EG (2014b) Soybean resistance to *Phakopsora pachyrhizi* as affected by acibenzolar-S-methyl, jasmonic acid and silicon. J Phytopathol 162:133–136. doi:10.1111/jph.12170

Curtis F, Cicco V, Lima G (2012) Efficacy of biocontrol yeasts combined with calcium silicate or sulphur for controlling durum wheat powdery mildew and increasing grain yield components. Field Crops Res 134:36–46. doi:10.1016/j.fcr.2012.04.014

Curvêlo CRS, Rodrigues FA, Silva LC, Nascimento KJT, Berger PG (2013) Mecanismos bioquímicos da defesa do algodoeiro à mancha de ramulária mediados pelo silício. Bragantia 72:41–51. doi:10.1590/S0006-87052013005000017

Dallagnol LJ, Rodrigues FA, Mielli MVB, Ma JF, Datnoff LE (2009) Defective active silicon uptake affects some components of rice resistance to brown spot. Phytopathology 99:116–121. doi:10.1590/S1982-56762013000600003

Dallagnol LJ, Rodrigues FA, Tanaka O, Amorim L, Camargo LEA (2012) Effect of potassium silicate on epidemic components of powdery mildew on melon. Plant Pathol 61:323–330. doi:10.1111/j.1365-3059.2011.02518.x

Dann EK, Muir S (2002) Peas grown in media with elevated plant-available silicon levels have higher activities of chitinases and β-1,3-glucanase, are less susceptible to a fungal leaf spot pathogen and accumulate more foliar silicon. Australas Plant Pathol 31:9–13

Datnoff LE, Nagata RT (1999) Influence of silicon on gray leaf spot development in St. Augustinegrass. Phytopathology 89:10

Datnoff LE, Rodrigues FA (2005) The role of silicon in suppressing rice diseases. APSnet Feature. doi:10.1094/APSnetFeature-2005-0205

Datnoff LE, Snyder GH (1994) Comparison of silicon and benomyl alone and in combination for reducing blast incidence. Biol Cult Tests Control Plant Dis 9:113

Datnoff LE, Raid RN, Snyder GH, Jones DB (1990) Evaluation of calcium silicate slag and nitrogen on brown spot, neck rot, and sheath blight development on rice. Biol Cult Tests Control Plant Dis 5:65

Datnoff LE, Snyder GH, Raid RN, Jones DB (1991) Effect of calcium silicate on blast and brown spot intensities and yields of rice. Plant Dis 75:729–732. doi:10.1094/PD-75-0729

Datnoff LE, Snyder GH, Deren CW (1992) Influence of silicon fertilizer grades on blast and brown spot development and on rice yields. Plant Dis 76:1011–1013. doi:10.1094/PD-76-1011

Datnoff LE, Deren CW, Snyder GH (1997) Silicon fertilization for disease management of rice in Florida. Crop Prot 16:525–531. doi:10.1016/S0261-2194(97)00033-1

Datnoff LE, Brecht MO, Stiles CM, Rutherford BA (2005) The role of silicon in suppressing foliar diseases in warm season turf. Int Turfgrass Soc Res J 10:175–179

Deepak S, Manjunath G, Manjula S, Niranjan-Raj S, Geetha NP, Shetty HS (2008) Involvement of silicon in pearl millet resistance to downy mildew disease and its interplay with cell wall proline/hydroxyproline-rich glycoproteins. Austral Plant Pathol 37:498–504. doi:10.1071/AP08047

Domiciano GP, Rodrigues FA, Moreira WR, Oliveira HV, Vale FXR, Filha MSX (2010a) Silício no progresso da mancha marrom na folha bandeira do trigo. Trop Plant Pathol 35:186–189. doi:10.1590/S1982-56762010000300009

Domiciano GP, Rodrigues FA, Vale FXR, Xavier Filha MS, Moreira WR, Andrade CCL, Pereira SC (2010b) Wheat resistance to spot blotch potentiated by silicon. J Phytopathol 158:334–343. doi:10.1111/j.1439-0434.2009.01623.x

Domiciano GP, Rodrigues FA, Guerra AMN, Vale FXR (2013) Infection process of *Bipolaris sorokiniana* on wheat leaves is affected by silicon. Trop Plant Pathol 38:258–263. doi:10.1590/S1982-56762013005000006

Duarte HSS, Zambolim L, Rodrigues FA (2007) Controle da requeima em tomateiro industrial com fungicidas e silicato de potássio. Fitopatol Bras 32:257–260. doi:10.1590/S0100-41582007000300012

Duarte HSS, Zambolim L, Rodrigues FA, Rios JA (2008) Efeito do silicato de potássio isoladamente ou em mistura com fungicida no controle da requeima da batateira. Summa Phytopathol 34:68–70. doi:10.1590/S1982-56762013005000006

Duarte HSS, Zambolim L, Rodrigues FA, Rios JA, Lopes UP (2009) Silicato de potássio, acibenzolar-S-metil e fungicidas no controle da ferrugem da soja. Cienc Rural 39:2271–2277. doi:10.1590/S0103-84782009000800001

Elawad SH, Green VE (1979) Silicon and the rice plant environment: a review of recent research. Riso 28:235–253

Feng XG, Chun SW, Lin SA, Jun LZ, Liang FF, Chao LY (2010) Influence of silicon on rice growth, resistance to bacterial blight and activity of pathogenesis-related proteins. Sci Agric Sin 43:690–697

Garibaldi A, Gilardi G, Gullino ML (2011) Silicon and electrical conductivity reducing powdery mildew of hydroponically grown tomato. Phytopathol Mediterr 50:192–202

Garibaldi A, Gilardi G, Cogliati EE, Gullino ML (2012) Silicon and increased electrical conductivity reduce downy mildew of soilless grown lettuce. Eur J Plant Pathol 132:123–132. doi:10.1111/j.1439-0434.2009.01665.x

Germar B (1934) Some functions of silicic acid in cereals with special reference to resistance to mildew. Zeitschrift Pflanzenemähr Düng Bodenk 35:102–115

Gória MM, Ghini R, Bettiol W (2013) Elevated atmospheric CO_2 concentration increases rice blast severity. Trop Plant Pathol 38:253–257. doi:10.1590/S1982-56762013005000010

Grothge-Lima MT, Lima Filho OF, Tsai SM (2001) A hydroponic survey on silicon, stem canker and root nodulation interactions in soybean (*Glycine max* Merril). In: Datnoff LE, Snyder GH, Korndörfer GH (eds) Silicon in agriculture, vol 8, Studies in plant science. Elsevier Science BV, Amsterdam, p 376

Guerra AMNM, Rodrigues FA, Berger PG, Barros AF, Rodrigues YC, Lima TC (2013a) Resistência do algodoeiro à ferrugem tropical potencializada pelo silício. Bragantia 72:279–291. doi:10.1590/brag.2013.037

Guerra AMNM, Rodrigues FA, Berger PG, Barros AF, Silva YCR, Lima TC (2013b) Aspectos bioquímicos da resistência do algodoeiro à ramulose potencializada pelo silício. Bragantia 72:292–303. doi:10.1590/brag.2013.036

Guével MH, Menzies JG, Bélanger RR (2007) Effect of root and foliar applications of soluble silicon on powdery mildew control and growth of wheat plants. Eur J Plant Pathol 119:429–436. doi:10.1007/s10658-007-9181-1

Hamel SC, Heckman JR (1999) Impact of mineral silicon products on powdery mildew in greenhouse grown turf, vol 31, Rutgers Turfgrass. Rutgers, New Jersey

Hashimoto A, Hirano K (1976) Statistical analysis of the integrated control measures against rice blast. Bull Fukushima Prefect Agric Exp Stn 15:95–108

Heckman JR, Johnston S, Cowgill W (2003) Pumpkin yield and disease response to amending soil with silicon. HortScience 38:552–554

Igarashi S (2008) Supressão de *Colletotrichum acutatum*, agente causal da flor-preta, na cultura do morangueiro, através da nutrição com cálcio (Ca) e silício (Si). Tese de doutorado em Agronomia-Universidade Estadual de Londrina, p 52

Jiang D (1993) The role of silicon in barley resistance to powdery mildew (*Erysiphe graminis* f. sp. *hordei*). PhD dissertation, University of Minnesota, St. Paul

Juliatti FC, Rodrigues FA, Korndörfer GH, Silva OA, Corrêa GF, Peixoto JR (1996) Effect of silicon in the induction of resistance to *Diaporthe phaseolorum* f. sp. *meridionalis* in soybean cultivars with different levels of resistance. Fitopatol Bras 21:362

Junior OS, Bonaldo SM (2013) Efeito de silício foliar no controle de brusone em arroz no norte de Mato Grosso. Sci Electron Arch 4:59–63

Kablan L, Lagauche A, Delvaux B, Legrève A (2012) Silicon reduces black sigatoka development in banana. Plant Dis 96:273–278. doi:10.1094/PDIS-04-11-0274

Kanto T (2002) Research of silicate for improvement of plant defense against pathogens in Japan. In: Matoh T (ed) Second silicon in agriculture conference. Press-Net, Kyoto, pp 22–26

Kanto T, Miyoshi A, Ogawa T, Maekawa K, Aino M (2004) Suppressive effect of potassium silicate on powdery mildew of strawberry in hydroponics. J Gen Plant Pathol 70:207–211. doi:10.1007/s10327-004-0117-8

Kanto T, Miyoshi A, Ogawa T, Maekawa T, Aino M (2006) Suppressive effect of liquid potassium silicate on powdery mildew of strawberry in soil. J Gen Plant Pathol 72:137–142. doi:10.1007/s10327-005-0270-8

Kanto T, Maekawa K, Aino M (2007) Suppression of conidial germination and appressorial formation by silicate treatment in powdery mildew of strawberry. J Gen Plant Pathol 73:1–7. doi:10.1007/s10327-006-0311-y

Kim SG, Kim KW, Park EW, Choi D (2002) Silicon-induced cell wall fortification of rice leaves: a possible cellular mechanism of enhanced host resistance to blast. Phytopathology 92:1095–1103. doi:10.1094/PHYTO.2002.92.10.1095

Kitani K, Natsume T, Asaga K (1960) Application of calcium silicate to control blast disease of rice. (II) Combined effect of calcium silicate and mercuric fungicide. Shikoku Agric Res 6:123–126

Kobayashi T, Ishiguro K, Nakajima T, Kim HY, Okada M, Kobayashi K (2006) Effects of elevated atmospheric CO_2 concentration on the infection of rice blast and sheath blight. Phytopathology 96:425–431. doi:10.1094/PHYTO-96-0425

Korndörfer GH, Datnoff LE, Corrêa GF (1999) Influence of silicon on grain discoloration and upland rice grown on four savanna soils from Brazil. J Plant Nutr 22:93–102. doi:10.1080/01904169909365609

Kozaka T (1965) Control of rice blast by cultural practices in Japan. In: The rice blast disease. Proceedings of a symposium at the International Rice Research Institute. Johns Hopkins University Press, Baltimore, pp 421–438

Kunoh H, Ishizaki H (1975) Silicon levels near penetration sites of fungi on wheat, barley, cucumber and morning glory leaves. Physiol Mol Plant Pathol 5:283–287. doi:10.1016/0048-4059(75)90094-6

Lemes E, Mackowiak CL, Blount A, Marois JJ, Wright DL, Coelho L, Datnoff LE (2011) Effects of silicon applications on soybean rust development under greenhouse and field conditions. Plant Dis 95:317–324. doi:10.1094/PDIS-07-10-0500

Leusch HJ, Buchenauer H (1988) Si-Gehalte und Si-Lokalisation im Weizenblatt und deren Bedeutung für die Abwehr einer Mehltauinfektion. Kali-Briefe (Büntehof) 19:13–24

Leusch HJ, Buchenauer H (1989) Effect of soil treatments with silica-rich lime fertilizers and sodium trisilicate on the incidence of wheat by *Erysiphe graminis* and *Septoria nodorum* depending on the form of N-fertilizer. J Plant Dis Plant Prot 96:154–172

Li W, Bi Y, Ge Y, Li Y, Wang J, Wang Y (2012) Effects of postharvest sodium silicate treatment on pink rot disease and oxidative stress-antioxidative system in muskmelon fruit. Eur Food Res Technol 234:137–145. doi:10.1007/s00217-011-1611-9

Liang YC, Sun WC, Si J, Römheld V (2005) Effects of foliar- and root-applied silicon on the enhancement of induced resistance to powdery mildew in *Cucumis sativus*. Plant Pathol 54:678–685. doi:10.1111/j.1365-3059.2005.01246.x

Lopes UP, Zambolim L, Lopes UN, Rios JA, Duarte HSS, Ribeiro Júnior JI (2013a) Silicate slag combined with tebuconazole in management of brown eye spot in coffee. Coffee Sci 8:221–226

Lopes UP, Zambolim L, Neto PNS, Souza AF, Capucho AS, Rodrigues FA (2013b) Effect of foliar application of potassium silicate on the progress of coffee leaf rust. Trop Plant Pathol 38:547–551. doi:10.1590/S1982-56762013000600012

Lopes UP, Zambolim L, Costa H, Pereira OL, Finger FL (2014a) Potassium silicate and chitosan application for gray mold management in strawberry during storage. Crop Prot 63:103–106. doi:10.1016/j.cropro.2014.05.013

Lopes UP, Zambolim L, Neto PNS, Duarte HSS, Ribeiro Júnior JI, Souza AF, Rodrigues FA (2014b) Silicon and triadimenol for the management of coffee leaf rust. J Phytopathol 162:124–128. doi:10.1111/jph.12166

Lu G, Jian W, Zhang J, Zhou Y, Cao J (2008) Suppressive effect of silicon nutrient on *Phomopsis* stem blight development in asparagus. HortScience 43:811–817

Mathai G, Paily PV, Menon MR (1977) Effect of fungicides and silica in the control of sheath blight disease of rice caused by *Corticium sasakii* (Shirai). Agric Res J Kerala 19:79–83

Menzies JG, Ehret DL, Glass ADM, Helmer T, Koch C, Seyward F (1991) Effects of soluble silicon on the parasitic fitness of *Sphaerotheca fuliginea* on *Cucumis sativus*. Phytopathology 81:84–88

Menzies J, Bowen P, Ehret DL, Glass ADM (1992) Foliar applications of potassium silicate reduce severity of powdery mildew on cucumber, muskmelon, and zucchini squash. J Am Soc Hortic Sci 117:902–905

Menzies JG, Ehret DL, Chérif M, Bélanger RR (2001) Plant-related silicon research in Canada. In: Datnoff LE, Snyder GH, Korndörfer GH (eds) Silicon in agriculture, vol 8, Studies in plant science. Elsevier Science BV, Amsterdam, pp 323–341

Miyaki Y, Takahashi E (1983) Effect of silicon on the growth of solution-cultured cucumber plants. Jpn J Soil Sci Plant Nutr 29:71–83

Moraes SRG, Pozza EA, Alves E, Pozza AAA, Carvalho JG, Lima PH, Botelho AO (2006) Efeito de fontes de silício na incidência e na severidade da antracnose do feijoeiro. Fitopatol Bras 31:69–75. doi:10.1590/S0100-41582006000100012

Moraes SRG, Pozza EA, Pozza AAA, Carvalho JG, Souza PE (2009) Nutrição do feijoeiro e intensidade da antracnose em função da aplicação de silício e cobre. Acta Sci Agron 31:283–291. doi:10.4025/actasciagron.v31i2.7037

Moyer C, Peres NA, Datnoff LE, Simonne EH, Deng Z (2008) Evaluation of silicon for managing powdery mildew on gerberadaisy. J Plant Nutr 31:2131–2144. doi:10.1080/01904160802459641

Muir S, Khoo C, McCabe B, Fensom G, Offord C, Brien J, Summerell B (2001) Some effects of silicon in potting mixes on growth and protection of plants against fungal diseases. In: Datnoff LE, Snyder GH, Korndörfer GH (eds) Silicon in agriculture, vol 8, Studies in plant science. Elsevier Science BV, Amsterdam, p 374

Naidoo PV, Caldwell PM, McFarlane S (2008) Movement of silicon through *Saccharum officinarum* (sugarcane) and its effect on *Puccinia melanocephala* (brown rust). In: 4th silicon in agriculture conference book of abstract, South Africa, p 79

Nakata Y, Ueno M, Kihara J, Ichii M, Taketa S, Arase S (2008) Rice blast disease and susceptibility to pests in a silicon uptake-deficient mutant *lsi1* of rice. Crop Prot 27:865–868. doi:10.1016/j.cropro.2007.08.016

Nanayakkara UN, Uddin W, Datnoff LE (2008) Effects of soil type, source of silicon, and rate of silicon source on development of gray leaf spot of perennial ryegrass turf. Plant Dis 92:870–877. doi:10.1094/PDIS-92-6-0870

Nanyakkara UN, Uddin W, Datnoff LE (2009) Soil silicon amendment for managing gray leaf spot of perennial ryegrass turf on golf courses in Pennsylvania. Can J Plant Pathol 31:1–12

Nanda HP, Gangopadhyay S (1984) Role of silicated cells in rice leaf on brown spot disease incidence by *Bipolaris oryzae*. Int J Trop Plant Dis 2:89–98

Narwal RP (1973) Silica bodies and resistance to infection in jowar (*Sorghum vulgare* Perc.). Agra Univ J Res Sci 22:17–20

O'Neill TM (1991) Investigation of glasshouse structure, growing medium and silicon nutrition as factors affecting disease incidence in cucumber crops. Med Fac Landbouww, Rijksuniv Gent 56:359–367

Ou SH (1985) Rice diseases, 2nd edn. Commonwealth Agricultural Bureau, Kew

Pagani APS, Dianese AC, Café-Filho AC (2014) Management of wheat blast with synthetic fungicides, partial resistance and silicate and phosphite minerals. Phytoparasitica 42:609–617. doi:10.1007/s12600-014-0401-x

Pereira SC, Rodrigues FA, Carré-Missio V, Oliveira MGA, Zambolim L (2009) Aplicação foliar de silício na resistência da soja à ferrugem e na atividade de enzimas de defesa. Trop Plant Pathol 34:164–170

Polanco LR, Rodrigues FA, Nascimento KJT, Shulman P, Silva LC, Neves FW, Vale FXR (2012) Biochemical aspects of bean resistance to anthracnose mediated by silicon. Ann Appl Biol 161:140–150. doi:10.1111/j.1744-7348.2012.00558.x

Polanco LR, Rodrigues FA, Moreira EN, Duarte HSS, Cacique IS, Valente LA, Vieira RF, Paula Júnior TJ, Vale FXR (2014) Management of anthracnose in common bean by foliar sprays of potassium silicate, sodium molybdate, and fungicide. Plant Dis 98:84–89. doi:10.1094/PDIS-03-13-0251-RE

Pozza AAA, Alves E, Pozza EA, Carvalho JG, Montanari M, Guimarães PTG, Santos DM (2004) Efeito do silício no controle da cercosporiose em três variedades de cafeeiro. Fitopatol Bras 29:185–188. doi:10.1590/s0100-41582004000200010

Prabhu AS, Filho MPB, Datnoff LE, Snyder GH, Berni RF, Rodrigues FA, Dallagnol LJ (2012) Silicon reduces brown spot severity and grain discoloration on several rice genotypes. Trop Plant Pathol 37:409–414. doi:10.1590/s1982-56762012000600005

Rabindra B, Gowda SS, Gowda KTP, Rajappa HK (1981) Blast disease as influenced by silicon in some rice varieties. Curr Res 10:82–83

Raid RN, Anderson DL, Ulloa MF (1992) Influence of cultivar and amendment of soil with calcium silicate slag on foliar disease development and yield of sugar-cane. Crop Prot 11:84–88

Ramos ARP, Santos RL, Amaro ACE, Fumes LAA, Boaro CSF, Cardoso AII (2013) Eficiência do silicato de potássio no controle do oídio e no desenvolvimento de abobrinha de moita. Hortic Bras 31:432–438

Rémus-Borel W, Menzies JG, Richard R, Bélanger RR (2005) Silicon induces antifungal compounds in powdery mildew-infected wheat. Physiol Mol Plant Pathol 66:108–115. doi:10.1016/j.pmpp.2005.05.006

Resende RS, Rodrigues FA, Soares JM, Casela CR (2009) Influence of silicon on some components of resistance to anthracnose in susceptible and resistant sorghum lines. Eur J Plant Pathol 124:533–541. doi:10.1007/s10658-009-9430-6

Resende RS, Rodrigues FA, Costa RV, Silva DD (2013) Silicon and fungicide effects on anthracnose in moderately resistant and susceptible sorghum lines. J Phytopathol 161:11–17. doi:10.1111/jph.12020

Rezende DC, Rodrigues FA, Carré-Missio V, Schurt DA, Kawamura K, Korndörfer GH (2009) Effect of root and foliar applications of silicon on brown spot development in rice. Australas Plant Pathol 38:67–73. doi:10.1071/ap08080

Rodgers-Gray BS, Shaw MW (2000) Substantial reductions in winter wheat diseases caused by addition of rice straw but not manure to soil. Plant Pathol 49:590–599

Rodgers-Gray BS, Shaw MW (2004) Effects of straw and silicon soil amendments on some foliar and stem-base diseases in pot-grown winter wheat. Plant Pathol 53:733–740. doi:10.1111/j.1365-3059.2004.01102.x

Rodrigues FA, Datnoff LE, Korndörfer GH, Seebold KW, Rush MC (2001) Effect of silicon and host resistance on sheath blight development in rice. Plant Dis 85:827–832. doi:10.1094/pdis.2001.85.8.827

Rodrigues FA, Vale FXR, Datnoff LE, Prabhu AS, Korndörfer GH (2003a) Effect of rice growth stages and silicon on sheath blight development. Phytopathology 93:256–261. doi:10.1094/phyto.2003.93.3.256

Rodrigues FA, Vale FXR, Korndörfer GH, Prabhu AS, Datnoff LE, Oliveira AMA, Zambolim L (2003b) Influence of silicon on sheath blight of rice in Brazil. Crop Prot 22:23–29. doi:10.1016/s0261-2194(02)00084-4

Rodrigues FA, Duarte HSS, Domiciano GP, Souza CA, Korndörfer GH, Zambolim L (2009) Foliar application of potassium silicate reduces the intensity of soybean rust. Australas Plant Pathol 38:366–372. doi:10.1071/ap09010

Rodrigues FA, Duarte HSS, Rezende DC, Wordell Filho JA, Korndörfer GH, Zambolim L (2010) Foliar spray of potassium silicate on the control of angular leaf spot on beans. J Plant Nutr 33:2082–2093. doi:10.1080/01904167.2010.519082

Rodrigues FA, Polanco LR, Duarte HSS, Resende RS, Vale FXR (2015) Photosynthetic gas exchange in common bean submitted to foliar sprays of potassium silicate, sodium molybdate and fungicide and infected with *Colletotrichum lindemuthianum*. J Phytopathol. doi:10.1111/jph.12353

Santos AB, Prabhu AS, Ferreira E, Fageria NK (2009) Fertilização silicatada na severidade de brusone e na incidência de insetos-praga em arroz irrigado. Rev Bras Eng Agric Ambient 13:537–543. doi:10.1590/S1415-43662009000500005

Santos GRS, Neto MDC, Ramos LN, Sarmento RA, Korndörfer GH, Ignácio M (2011) Effect of silicon sources on rice diseases and yield in the State of Tocantins, Brazil. Maringá 33:451–456. doi:10.4025/actasciagron.v33i3.6573

Savvas D, Giotis D, Chatzieustratiou E, Bakea M, Patakioutas G (2009) Silicon supply in soilless cultivations of zucchini alleviates stress induced by salinity and powdery mildew infections. Environ Exp Bot 65:11–17. doi:10.1016/j.envexpbot.2008.07.004

Schuerger AC, Hammer W (2002) Suppression of powdery mildew on greenhouse-grown cucumber by addition of silicon to hydroponic nutrient solution is inhibited at high temperature. Plant Dis 87:177–185. doi:10.1094/PDIS.2003.87.2.177

Schultz B, Bora KC, Nogueira AC, Auer CG (2012) Uso do silicato de potássio no controle de oídio em mudas de *Eucalyptus benthamii*. Pesq Florest Bras 32:93–100. doi:10.4336/2012.pfb.32.69.93

Seebold KW, Datnoff LE, Correa-Victoria FJ, Kucharek TA, Snyder GH (2000) Effect of silicon rate and host resistance on blast, scald, and yield of upland rice. Plant Dis 84:871–876. doi:10.1094/PDIS.2000.84.8.871

Seebold KW, Kucharek TA, Datnoff LE, Correa-Victoria FJ, Marchetti MA (2001) The influence of silicon on components of resistance to blast in susceptible, partially resistant, and resistant cultivars of rice. Phytopathology 91:63–69

Seebold KW, Datnoff LE, Correa-Victoria FJ, Kucharek TA, Snyder GH (2004) Effects of silicon and fungicides on the control of leaf and neck blast in upland rice. Plant Dis 88:253–258. doi:10.1094/PDIS.2004.88.3.253

Shetty R, Jensen B, Shetty BP, Hansen M, Hansen CW, Starkey KR, Jørgensen HJL (2012) Silicon induced resistance against powdery mildew of roses caused by *Podosphaera pannosa*. Plant Pathol 61:120–131. doi:10.1111/j.1365-3059.2011.02493.x

Silva IT, Rodrigues FA, Oliveira JR, Pereira SC, Andrade CCL, Silveira PR, Conceição MM (2010) Wheat resistance to bacterial leaf streak mediated by silicon. J Phytopathol 158:253–262. doi:10.1111/j.1439-0434.2009.01610.x

Soratto RP, Fernandes AM, Crusciol CAC, Sousa-Schlick GDS (2012) Produtividade, qualidade de tubérculos e incidência de doenças em batata, influenciados pela aplicação foliar de silício. Pesq Agrop Brasileira 47:1000–1006. doi:10.1111/j.1365-3059.2009.02228.x

Sun W, Zhang J, Fan Q, Xue G, Li Z, Liang Y (2010) Silicon-enhanced resistance to rice blast is attributed to silicon-mediated defence resistance and its role as physical barrier. Eur J Plant Pathol 128:39–49. doi:10.1007/s10658-010-9625-x

Suzuki H (1935) The influence of some environmental factors on the susceptibility of the rice plant to blast and Helminthosporium diseases and on the anatomical characters of the plant III. Influence of differences in soil moisture and in amounts of fertilizer and silica given. J Coll Agric 13:277–331

Takahashi Y (1967) Nutritional studies on development of Helminthosporium leaf spot. In: Proceedings of the symposium on rice diseases and their control by growing resistant varieties and other measures. Agriculture, Forestry, and Fisheries Research Council, Tokyo, pp 157–170

Tamimi YN, Hunter JE (1970) Effect of P, $CaCO_3$ and $CaSiO_3$ fertilization upon incidence of corn smut. Agron J 62:496–498

Tatagiba SD, Rodrigues FA, Filippi MCC, Silva GB, Silva LC (2014) Physiological responses of rice plants supplied with silicon to *Monographella albescens* infection. J Phytopathol 162:596–606. doi:10.1111/jph.12231

Tesfagiorgis HB, Laing MD (2011) Effects of concentration, frequency of application and runoff of foliar-applied soluble silicon on powdery mildew of zucchini. Afr J Agric Res 6:2243–2248

Tesfagiorgis HB, Laing MD, Annegarn HJ (2014) Evaluation of biocontrol agents and potassium silicate for the management of powdery mildew of zucchini. Biol Control 73:8–15. doi:10.1016/j.biocontrol.2014.03.010

Volk RJ, Kahn RP, Weintraub RL (1958) Silicon content of the rice plant as a factor in influencing its resistance to infection by the rice blast fungus, *Piricularia oryzae*. Phytopathology 48:121–178

Voogt W (1992) The effects of Si application on roses in rockwool. In: Annual Report Glasshouse Crops Research Station, Naaldwijk, pp 17–18

Voogt W, Sonneveld C (2001) Silicon in horticultural crops grown in soilless culture. In: Datnoff LE, Snyder GH, Korndörfer GH (eds) Silicon in agriculture, vol 8, Studies in plant science. Elsevier Science BV, Amsterdam, pp 115–131

Wagner F (1940) Die Bedeutung der Kieselsäure für das Wachstum einiger Kulturpflanzen, ihren Närstoffhaushalt und ihre Anfälligkeit gegen echte Mehltaupilze. Phytopathol Z 12:427–479

Wiese J, Wiese H, Schwartz J, Schubert S (2005) Osmotic stress and silicon act additively in enhancing pathogen resistance in barley against barley powdery mildew. J Plant Nutr Soil Sci 168:269–274. doi:10.1002/jpln.200420490

Winslow MD (1992) Silicon, disease resistance, and yield of rice genotypes under upland cultural conditions. Crop Sci 32:1208–1213

Wolff SA, Karoliussen I, Rohloff J, Strimbec R (2012) Foliar application of silicon fertilizers inhibit powdery mildew development in greenhouse cucumber. J Food Agric Environ 10:355–359

Wordell Filho JA, Duarte HSS, Rodrigues FA (2013) Efeito da aplicação foliar de silicato de potássio e de fungicida na severidade da ferrugem da folha e da mancha amarela do trigo. Rev Ceres 60:726–730. doi:10.1590/S0034-737X2013000500018

Xavier Filha MS, Rodrigues FA, Domiciano GP, Oliveira HV, Silveira PR, Moreira WR (2011) Wheat resistance to leaf blast mediated by silicon. Australas Plant Pathol 40:28–38. doi:10.1007/s13313-010-0010-1

Yamauchi M, Winslow MD (1987) Silica reduces disease on upland rice in a high rainfall area. Int Rice Res Newslett 12:22–23

Yamauchi M, Winslow MD (1989) Effect of silica and magnesium on yield of upland rice in the humid tropics. Plant and Soil 113:265–269

Yanar Y, Yanar D, Gebologlu N (2011) Control of powdery mildew (*Leveillula taurica*) on tomato by foliar sprays of liquid potassium silicate (K_2SiO_3). Afr J Biotechnol 10:3121–3123. doi:10.5897/AJB11.215

Yang L, Zhao P, Wang L, Filippus I, Meng X (2010) Synergistic effect of oligochitosan and silicon on inhibition of *Monilinia fructicola* infections. J Sci Food Agric 90:630–634. doi:10.1002/jsfa.3860

Zanão Junior LA, Rodrigues FA, Fontes RLF, Korndörfer GH, Neves JCL (2009) Rice resistance to brown spot mediated by silicon and its interaction with manganese. J Phytopathol 157:73–78. doi:10.1111/j.1439-0434.2008.01447.x

Zanão Júnior LA, Fontes RLF, Coelho PHM, Korndörfer GH, Zambolim L (2010) Soil-applied silicon decreases severity of wheat spot blotch on silicon-deficient soils. Rev Bras Ciênc Solo 34:401–408. doi:10.1590/S0100-06832010000200013

Zellner W, Frantz J, Leisner S (2011) Silicon delays tobacco ringspot virus systemic symptoms in *Nicotiana tabacum*. J Plant Physiol 168:1866–1869. doi:10.1016/j.jplph.2011.04.002

Chapter 5
Silicon Potentiates Host Defense Mechanisms Against Infection by Plant Pathogens

Fabrício A. Rodrigues, Renata Sousa Resende, Leandro José Dallagnol, and Lawrence E. Datnoff

Abstract Several agronomic and horticultural crops, such as barley, cucumbers, oats, rice, sugarcane, and wheat, benefit from applications of silicon. Growth enhancements results, in part, from reductions in the intensities of plant diseases. For the rice-*Pyricularia oryzae* model pathosystem, the mechanical barrier formed from silicon polymerization below the cuticle and in the cell walls was the first proposed hypothesis to explain how this element reduced the number of blast lesions and the lesion sizes. However, new insights have revealed that silicon's effect on plant resistance to a number of diseases may also occur through mediated host plant resistance mechanisms against pathogen infection. Plants supplied with silicon exhibit potentiated activation of the phenylpropanoid pathway resulting in increases in total soluble phenolics and lignin. The activities of defense enzymes, such as chitinases and β-1,3-glucanases, are maintained at higher levels during infection and the transcription of defense related genes occur faster and with greater output. When plants are supplied with silicon and then challenged with a pathogen, there is an enhanced activation in antioxidant metabolism, which in turn, suppresses the damaging cytotoxic effect of the reactive oxygen species that causes lipid peroxidation in the cell membrane. At the physiological level, leaf gas exchange parameters of silicon-treated plants are higher upon pathogen infection for crops, such as common beans, rice, sorghum and wheat, indicating the ameliorating effect of this element on photosynthesis. Although our understanding of how silicon affects plants in response to infection has advanced, the exact mechanism(s) by which silicon

F.A. Rodrigues (✉) • R.S. Resende
Departamento de Fitopatologia, Universidade Federal de Viçosa, Viçosa, Minas Gerais Cep 36570-900, Brazil
e-mail: fabricio.plantpathologist@gmail.com; fabricio@ufv.br

L.J. Dallagnol
Departamento de Fitossanidade, Universidade Federal de Pelotas, Pelotas, Rio Grande do Sul Cep 96010-900, Brazil

L.E. Datnoff
Department of Plant Pathology and Crop Physiology, Louisiana State University Agricultural Center, 302 Life Sciences, Baton Rouge, LA 70803, USA
e-mail: LDatnoff@agcenter.lsu.edu

F.A. Rodrigues, L.E. Datnoff (eds.), *Silicon and Plant Diseases*,
DOI 10.1007/978-3-319-22930-0_5

modulates plant physiology through the potentiation of host defense mechanisms still requires further investigation at the genomics, proteomics, and metabolomics levels.

Introduction

The beneficial effects of silicon in plants under biotic and/or abiotic stresses, whether direct or indirect, reportedly occur in a wide variety of crops such as barley, cucumbers, oats, rice, sugarcane and wheat. Although the most remarkable effect of silicon is the reduced intensity of a number of plant diseases in many crops of great economic importance, the hypothesis that was first proposed for this underlying phenomenon was a mechanical barrier resulting from silicon polymerization in plant tissue. However, in addition to this passive mechanical role played by silicon, a plethora of biological, physiological and molecular data now suggests that this element may act as a modulator of host resistance against pathogen infection. To gain further understanding of this subject, we will discuss the mechanisms involved in host resistance against pathogen infections as mediated by silicon as well as highlights of our current knowledge at the -omics level in this chapter.

The Physical Barrier Hypothesis

To establish a successful infection, plant pathogens must gain access to the host's tissue by overcoming the physical barriers conferred primarily by wax, the cuticle and a thick cell wall (Freeman and Beattie 2008). The microscopic evidence used to explain how silicon may increase host resistance against plant diseases is based on the following: pre-formed defense barriers (a cuticle and cell wall) and post-formed defense barriers (papillae deposition and cell wall reinforcement at the infection sites) in an attempt to avoid or delay pathogen ingress.

The physical barrier was the first mechanism proposed to explain why silicon increased rice resistance to blast caused by *Pyricularia oryzae*. The great number of silicified bulliform cells in the epidermis of rice leaves was believed to act as a physical barrier that efficiently impeded or delayed penetration by *P. oryzae* (Ito and Hayashi 1931; Suzuki 1940; Hemmi et al. 1941). This physical barrier hypothesis gained more credence because a silica layer with a thickness of approximately 2.5 μm was observed beneath the cuticle of rice leaves and sheaths (Yoshida et al. 1962). This cuticle-silicon double layer was associated with a decrease in the number of blast lesions observed on the leaf blades and with a reduction in the number of infection pegs formed by the appressoria that pierced the underlying cell wall, allowing fungal access into the epidermal cell (Yoshida et al. 1962). Indeed, more detailed studies showed that the epidermal cell wall of plants that were supplied with silicon was made of an outer electron-dense silicon layer and an inner electron-translucent layer, which often had thin, electron-dense silicon layers embedded in cellulose microfibrils (Kim et al. 2002). Interestingly, the epidermal cell wall thickness was not significantly affected by silicon. However, the thickness ratios of silica layers to epidermal cell walls were much higher for plants from a resistant cultivar than for plants from a susceptible one. This finding supported the idea that silicified epidermal cell walls were closely associated with reduced blast severity for plants that were supplied with silicon (Kim et al. 2002). He and colleagues (2015) recently reported that most silicon was cross-linked with hemicellulose in the rice cell wall, which improved both the mechanical properties and the regeneration of the cell walls.

Seebold et al. (2001) noted that the reduced number of blast lesions (which were evaluated as the relative infection efficiency) on rice leaves from partially resistant and susceptible cultivars that were amended with silicon had fewer successful established infections per unit of inoculum, lending partial support to the physical barrier hypothesis. The reduction in the number of blast lesions as the silicon rates increased in the soil clearly indicated that silicon manifested its beneficial effects before the penetration peg from *P. oryzae* gained full access to the epidermal cell. Therefore, Seebold et al. (2001) proposed that silicon does more than just act as a physical barrier in rice resistance against blast. Based on light microscopy observations of the leaf adaxial surfaces from rice plants that were supplied with silicon, Hayasaka et al. (2008) noticed that the number of appressorial sites for *P. oryzae* with successful penetration was reduced in proportion to the amount of silicon deposited in the leaf epidermis. Although this fact does not necessarily support a cause-and-effect relation between the denser silicon layer and the reduced number of appressorial sites for *P. oryzae* with successful penetration, it is plausible that the denser silicon layer contributed to a longer incubation period. These studies emphasized the importance of the silicon deposition beneath the rice cuticle and in the cell wall to prevent or delay penetration by *P. oryzae*. Abed-Ashtiani et al. (2012) also observed that the blast severity dramatically decreased as the foliar silicon concentration increased when increasing silicon rates were added to the soil. In an oat-*Blumeria graminis* f. sp. *avenae* interaction, the fortification of the epidermal cell walls through silicification was also reported as a structural barrier against fungal penetration (Carver et al. 1998).

Although the silicification of the epidermal cell walls was believed to be the primary cause associated with the reduced number of leaf blast lesions, no direct supportive evidence was provided to show that the narrow fungal penetration peg did not actually overcome the cuticle-silicon double layer and the epidermal cell silicification. For many years, the density of silicified cells in the leaf epidermis of some rice cultivars was known for not being proportional to their level of blast resistance always (Hashioka 1942; Kawamura and Ono 1948). This effect may be related to the fact that the silicified cells, in which silicon was deposited and polymerized in the form of amorphous silica bodies, were not uniformly distributed throughout the leaf surfaces, thereby leaving unprotected areas of the leaf surface exposed to pathogen penetration (Kim et al. 2002; Motomura et al. 2004; Ma and Yamaji 2006;

Cacique et al. 2013). Rodrigues and his colleagues (2005) noted a decrease in the number of leaf blast lesions in rice plants that were supplied with silicon, which was likely caused by the inability of the *P. oryzae*-formed appressoria to overcome the physical impediment created by the cuticle-silica double layer. However, the presence of silica cells and silica bodies were again observed to be not uniformly distributed in the adaxial epidermis of leaves, and, as a consequence, they may have allowed successful fungal penetration at some infection sites.

It is known that the resistance of epidermal cells against fungal penetration is not strictly related to the increased thickness of the cuticle-silicon double layer or the number of silica cells found in the leaf epidermis (Rodrigues et al. 2005). Studies measuring the puncture resistance of rice epidermal cells to a needle tip from beneath a torsion balance in leaves were collected from rice plants that had been supplied with different silicon rates, and the results suggest that the puncture resistance was not explained solely by leaf epidermis silicification (Ishiguro 2001). Rather, this resistance might also be attributed to the nature of the epidermal cell protoplasm (Ito and Sakamoto 1939). Schurt et al. (2012) also observed that the leaf sheaths of rice plants that were supplied with silicon had increased puncture resistance. Therefore, the high silicon deposition most likely contributed to a delay in leaf sheath colonization by *Rhizoctonia solani*.

Domiciano et al. (2013) noted that the time needed for *Bipolaris sorokiniana* ingress into wheat epidermal cells was lengthened, and the foliar tissue colonized by the fungus was reduced because of the physical barrier formed by the double cuticle-silicon layer. According to these authors, this physical barrier may have reduced the diffusion of the lytic enzymes and the non-host selective toxins released by the pathogen at the leaf surface as shown by the reduced degradation of the waxy layer. Sousa et al. (2013) investigated the effect of silicon on cytological aspects arising from the infection of wheat leaves by *P. oryzae* at the microscopic level. According to these authors, *P. oryzae* hyphae grew successfully and formed an extensively branched mycelium in the first-invaded epidermal cell and then invaded several neighboring leaf cells from plants that were not supplied with silicon. By contrast, the leaves of silicon-supplied plants contained fungal hyphae that were restricted to the first-invaded epidermal cell. The number of brown (necrotic) adaxial epidermal cells and their browning intensities were significantly lower for silicon-supplied plants than those that were not supplied with silicon. The frequency of appressorial sites that exhibited a type B reaction (infection hyphae within the epidermal cell and an absence of cytoplasm granulation) was lower for silicon-supplied plants than for those that were not supplied from 72 to 96 h after inoculation, and the frequency of appressorial sites showing a type A reaction (unsuccessful penetration) was much higher in comparison with the non-supplied plants as well. Schurt et al. (2015) used light microscopy and scanning electron microscopy to observe the reduced growth of *R. solani* on the leaf sheaths of rice plants that were supplied with silicon, which exhibited intense autofluorescence in tissues near necrotic areas because of fungal colonization.

In addition to the reinforcement of cell walls by silicon, the formation of papillae has also been greatly stimulated by this element. Carver et al. (1987) observed localized silicon deposition in host cells beneath the appressoria of *B. graminis* f. sp. *hordei*, which failed to penetrate the barley epidermal cells. Silicon accumulation was found to occur in the haustorial neck and collar area of the fungus as well as in the papillae, regardless of the outcome of attempted penetration. At 20 h after inoculation, when the successful and failed penetration attempts became evident, the silicon concentration was three to four times greater at the infection sites where fungal penetration failed in comparison with the infection sites where successful fungal penetration occurred. The absence of high background silicon levels in the barley epidermal cells that were distant and adjacent to the penetrated cells suggested that the cuticle-silicon double layer likely did not play a role in the increasing barley resistance to powdery mildew, unlike the findings reported earlier in rice epidermal cells against *P. oryzae* (Kim et al. 2002). This finding further suggests that silicon was deposited in response to the penetration of the barley epidermal cells by *B. graminis* f. sp. *hordei* (Carver et al. 1998). Zeyen et al. (1993) demonstrated that barley epidermal cells would produce papillae in response to infections by *B. graminis* f. sp. *hordei* in the presence of soluble silicon. This finding suggested that an active process occurs in the cytoplasmic aggregate that then presumably concentrated soluble silicon and prevented its polymerization before it was transported across the plasma membrane into the epidermal cell wall and mature papillae. Jiang (1993) experimentally interrupted the papilla deposition in barley leaf epidermal cells that were and were not supplied with silicon, and were inoculated with *B. graminis* f. sp. *hordei*. The researchers noted a delay in the fungal penetration of leaves from silicon-supplied plants before papilla formation. Because soluble silicon was abundant at that time, it is unlikely that the physical barrier that was formed by insoluble silicon was more important in increasing resistance to penetration. Furthermore, the deposition of silicon appears to have required the availability of phenolics and hydroxyproline-rich glycoproteins in the leaf epidermal cells to prevent haustorium formation by *Uromyces vignae* in the leaves of French beans (Perumalla and Heath 1991). According to these authors, although the callose and cell wall reinforcement contributed to preventing haustorium formation, silicon deposition played the most pivotal role in this process. Kauss et al. (2003) reported that a strongly cationic, proline-rich protein reinforced the cell wall at the infection sites, and silicon deposition was enhanced during the development of systemic acquired resistance in cucumber leaves in response to *Colletotrichum lagenarium* infection, thus preventing fungal infection. In roses, the quantity of papillae was greater in the leaf cells of plants that were supplied with silicon in response to *Podosphaera pannosa* infection (Shetty et al. 2012). In wheat, the epidermal cells of plants that were supplied with silicon reacted against *B. graminis* f. sp. *tritici* infection by massive papilla formation (Bélanger et al. 2003). Pozza et al. (2004) reported a thicker cuticle on the lower leaf surface of coffee seedlings that received silicon. This thickened cuticle helped to reduce the penetration of *Cercospora coffeicola* and subsequently reduced the number of leaf lesions that developed. Taken together, these observations indicate that the proposed physical silicon barrier enhances resistance by decreasing the intensity of a number of plant diseases, and the mechanism is likely very complex.

The contribution of the physical barrier to the mechanism of silicon-conferred resistance against plant diseases (as conferred by silicon polymerization beneath the cuticle and in the cell wall) is still not widely agreed on. According to Fauteux et al. (2005), a silicon accumulation as determined by scanning electron microscopy and X-ray microanalysis at the leaf infection sites of *Arabidopsis thaliana* by *Erysiphe cichoracearum* was attributed to higher transpiration rates caused by cuticle damage rather than active transport. According to Samuels et al. (1991), cucumber plants that were transferred from pots containing 100 ppm of silicon to pots without silicon had higher powdery mildew severities. According to these authors, the presence of polymerized silicon in the cucumber leaves before the inoculum arrival did not seem to be more important than the constant presence of soluble silicon during the time course of the fungal infection. Sun et al. (2010) showed that rice plants that were initially grown in the presence of silicon and were then switched to a nutrient solution without this element prior to inoculation with *P. oryzae* still exhibited lower blast severity. However, the disease severity levels were greater for this treatment when compared with those of inoculated plants that received a continuous supply of silicon. It is likely that insoluble silicon affected blast development by physically strengthening the cell wall and thus reduced fungal leaf colonization (Sun et al. 2010). However, the authors also highlighted that the deposited (insoluble) silicon may not be as important as the available silicon (soluble) that is found in the cells at the time of fungal infection for reducing disease severity (Sun et al. 2010). These findings help to support the concept that a reduction in disease intensities cannot solely be attributed to the presence of insoluble silicon in the papillae and cell wall as reported for cucumber epidermal cells. Furthermore, Chérif et al. (1992b) observed the deposition of silicon in needle-punctured leaf holes and the absence of these deposits when the plants were grown under saturated humidity. This finding again suggested that when silicon accumulated in areas in which the cuticle was damaged, this accumulation was caused by an increase in the transpiration rate. Menzies et al. (1991) reported a negative correlation between a high foliar silicon concentration in cucumber plants with the leaf area covered by *Sphaerotheca fuliginea* colonies, the number of colonies per leaf, the individual colony size and the germination of conidia. These authors believed that the increased resistance of cucumber leaves to powdery mildew was associated with the reinforcement of epidermal cell walls by silicon. Samuels et al. (1991) also noted that the accumulation of silicon around powdery mildew colonies on cucumber leaves affected the fungal growth and, consequently, the diameter of the colonies.

Taken together, these studies clearly showed that silicon is deposited below the cuticle and in the cell walls, and it contributed in part to increased physical resistance against pathogen penetration. However, insoluble silicon may not be more important than soluble silicon in enhancing resistance to infection by plant pathogens. The resistance of plants that were supplied with silicon to both soil-borne and foliar diseases is a very complex phenomenon, and the physical resistance barrier is likely only one small aspect of how silicon confers plant disease resistance.

Biochemical and Molecular Aspects of Silicon-Mediated Host Resistance to Pathogens

Wheat plants that were supplied with silicon produced fungitoxic aglycones in response to *Blumeria graminis* f. sp. *tritici* infection as demonstrated by TLC chromatogram analyses coupled with bioassays (Rémus-Borel et al. 2005). According to these authors, after the leaf fractions were analyzed by high-performance liquid chromatography and comparative analyses of the profiles were performed, at least three compounds were confirmed to occur in higher amounts for inoculated wheat plants that had been supplied with silicon (Rémus-Borel et al. 2005). Fresh transverse sections of leaves from wheat plants that were supplied with silicon and infected with *B. graminis* f. sp. *tritici* were analyzed by fluorescence microscopy, and intense autofluorescence was observed. This finding suggested that the presence of phenolics likely contributed to the collapse of conidial chains at the examined fungal infection sites (Rémus-Borel et al. 2005). Rémus-Borel et al. (2009) further investigated if trans-aconitate (TA) could act as a precursor of methylated TA forms in wheat, and they addressed the possible relations between the silicon supply, disease development, and TA and methyl TA concentrations in leaf tissues. According to these authors, the TA concentration in non-inoculated plants increased as the disease progressed, regardless of the presence of silicon. By contrast, the TA concentration remained fairly constant in the leaf tissue of inoculated plants regardless of whether silicon was present. However, for plants that were supplied with silicon, the TA concentration was significantly lower than it was for the plants that were not supplied with silicon. For the inoculated plants supplied with silicon, an increase in wheat resistance to powdery mildew was closely associated with the methyl TA concentration. Silicon apparently had an effect on the methyl TA concentration only for inoculated plants, suggesting that this element does not act directly on the TA concentration, but increases the production of methyl TA for infected plants. Based on the increase in methyl TA and the leveling off of the TA concentration, it appears that the latter, instead of accumulating, was used by the diseased plants to produce methylated forms of antifungal TA so that they would act as phytoalexins to decrease disease development. This observed phenomenon was more pronounced for silicon-supplied plants (Rémus-Borel et al. 2009).

Rodrigues et al. (2003) provided the first cytological evidence that silicon-mediated resistance to *P. oryzae* in rice was correlated with a specific leaf cell reaction that interfered with pathogen development. Ultrastructural observations of samples from plants that were not supplied with silicon revealed that some host cells were devoid of organelles and that some host cell walls were no longer discernible in the massively colonized mesophyll and vascular bundle (Fig. 5.1a and b). A light deposition of osmiophilic material with a granular texture that occasionally interacted with fungal walls was observed in some epidermal cells (Fig. 5.1c, arrows). In plants that received silicon, empty fungal hyphae were evenly surrounded by a dense layer of granular osmiophilic material that partially occluded the epidermal cells (Fig. 5.1d, arrows), the vascular bundle (Fig. 5.1e, arrowheads)

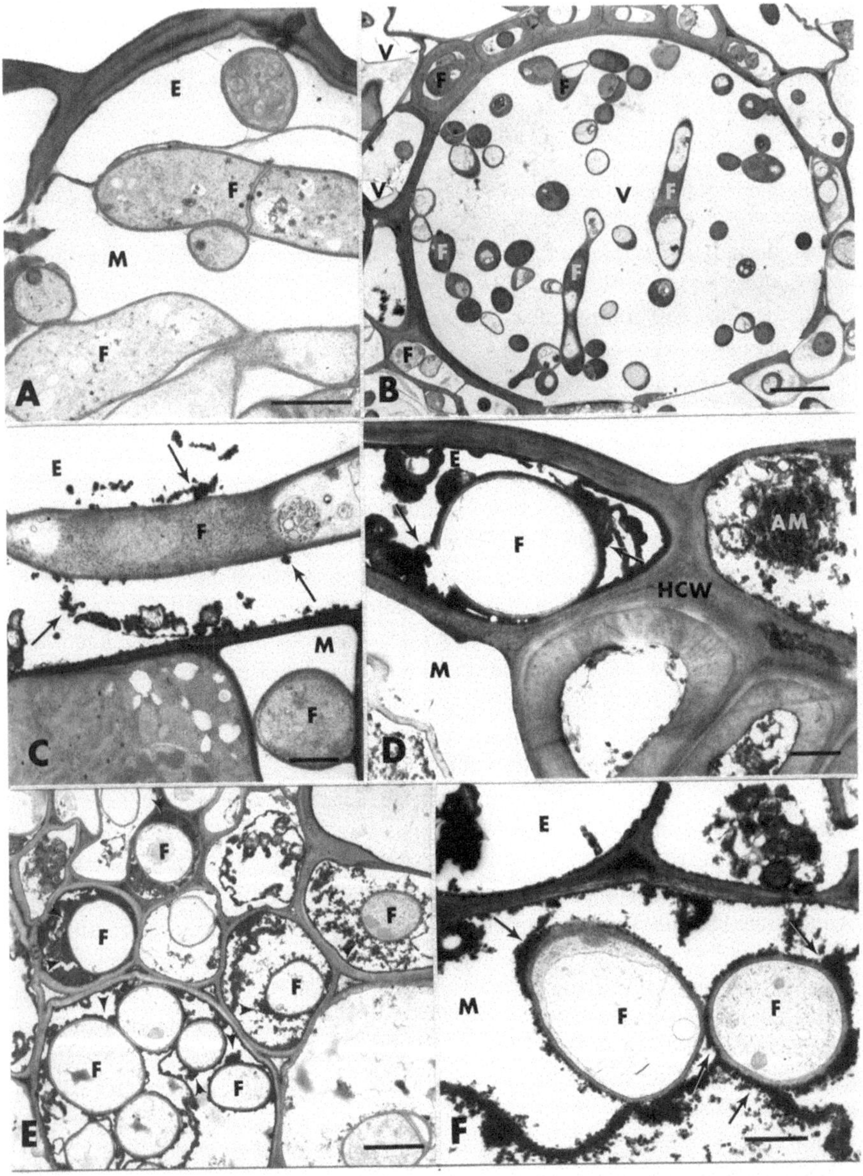

Fig. 5.1 Transmission electron micrographs of leaf samples collected from rice plants non-supplied (−Si) and supplied (+Si) with silicon (Si) at 96 h after inoculation with *Pyricularia oryzae*. (**a**) Ultrastructurally normal fungal hyphae colonize both the epidermis and mesophyll. Host cell walls are no longer discernible in the mesophyll (−Si). Bar = 2 μm. (**b**) The vascular bundle is massively colonized by the fungal hyphae (−Si). Bar = 5 μm. (**c**) Some amorphous material (*arrows*) accumulates in an epidermal cell and irregularly interacts with a fungal cell wall (−Si). Bar = 1 μm. (**d**) A dense amorphous material (*arrows*) accumulates around an empty fungal hyphae in the epidermal cell and also is found in an epidermal cell neighboring the colonized one (+Si). Bar = 1 μm. (**e**) Fungal hyphae invading the vascular bundle are often surrounded by dense amorphous material and often reduced to empty shells (+Si; *arrowheads*). Bar = 2 μm. (**f**) Two fungal hyphae in a mesophyll cell are evenly coated by the amorphous material (+Si; *arrows*) Bar = 1 μm. *AM* Amorphous material, *F* fungal hyphae, *E* epidermis, *M* mesophyll, *HCW* host cell wall, and *V* vascular bundle (Reproduced from Datnoff and Rodrigues (2005))

and the mesophyll cells (Fig. 5.1f, arrows). The cytochemical labeling of chitin revealed no difference in the pattern of chitin localization over fungal cell walls regardless of the presence of silicon at 96 h after inoculation, indicating a limited production of chitinases by the rice plant as a mechanism of defense. However, the occurrence of empty fungal hyphae that were surrounded or trapped in amorphous material, which were found in samples from plants that were supplied with silicon, suggested that phenolic-like compounds or phytoalexins played a primary role in rice defense response against *P. oryzae* infection. In a further study, Rodrigues et al. (2004) provided evidence that higher levels of momilactone phytoalexins were found in leaf extracts from plants that were inoculated with *P. oryzae* and supplied with silicon than in leaf extracts from inoculated plants that were not supplied with silicon or were not inoculated and supplied with silicon. On this basis, the more efficient terpenoid pathway stimulation in the plants receiving silicon and, consequently, the increase in the momilactone levels, appeared to be a factor that contributed to enhanced rice resistance to blast. Maekawa et al. (2002) observed a dramatic increase in superoxide generation in the rice leaves of plants that were supplied with silicon 15 min after *P. oryzae* inoculation. After this time, the superoxide generation rapidly decreased to levels observed for inoculated plants that had not been supplied with silicon. Fortunato et al. (2014) performed a study to investigate, at the histochemical level, whether silicon could enhance the production of phenolics in banana roots in response to *F. oxysporum* f. sp. *cubense* infection. According to these authors, intense orange-yellow autofluorescence was detected in the metaxylem and phloem vessels of the root sections of inoculated plants that were not supplied with silicon at 24 and 32 dai, respectively (Figs. 5.2a and 5.3a). Autofluorescence was also observed in the phloem vessels and the sclerenchyma cells in the root sections of the inoculated plants that received silicon at 24 and 32 dai (Figs. 5.2b and 5.3b). For non-inoculated plants, the autofluorescence in the medulla and in the cortex was weak regardless of whether silicon was provided. There was an absence of fluorescence in the root sections of the inoculated plants that were not supplied with silicon when they were treated with both Neu's and Wilson's reagents, which are used to stain flavonoid compounds, at 24 and 32 dai (Figs. 5.2c and e; 5.3c and e). At 24 dai, a strong yellow-orange fluorescence was observed in the phloem and a lemon-yellow fluorescence was observed in the sclerenchyma and metaxylem vessels in the root sections of the inoculated plants that had been supplied with silicon stained with Neu's reagent (Figs. 5.2d and 5.3d). With Wilson's reagent, an orange-yellow autofluorescence was more pronounced around the phloem vessels, and a yellow fluorescence was more pronounced around the metaxylem vessels in the root sections of the inoculated plants that received silicon at 24 and 32 dai (Figs. 5.2f and 5.3f). Lignin was densely deposited in the cortex of the root sections in the inoculated plants that received silicon (Figs. 5.2h and 5.3h) in comparison with the root sections of the inoculated plants that did not at 24 dai (Fig. 5.2g) and 32 dai (Fig. 5.3g). Dopamine was barely detected in the root sections of the inoculated plants that were not supplied with silicon at 24 and 32 dai according to lactic and glyoxylic acid stains (Figs. 5.2i and 5.3i). However, dopamine was strongly suspected to occur in the phloem and metaxylem vessels of the root sections in

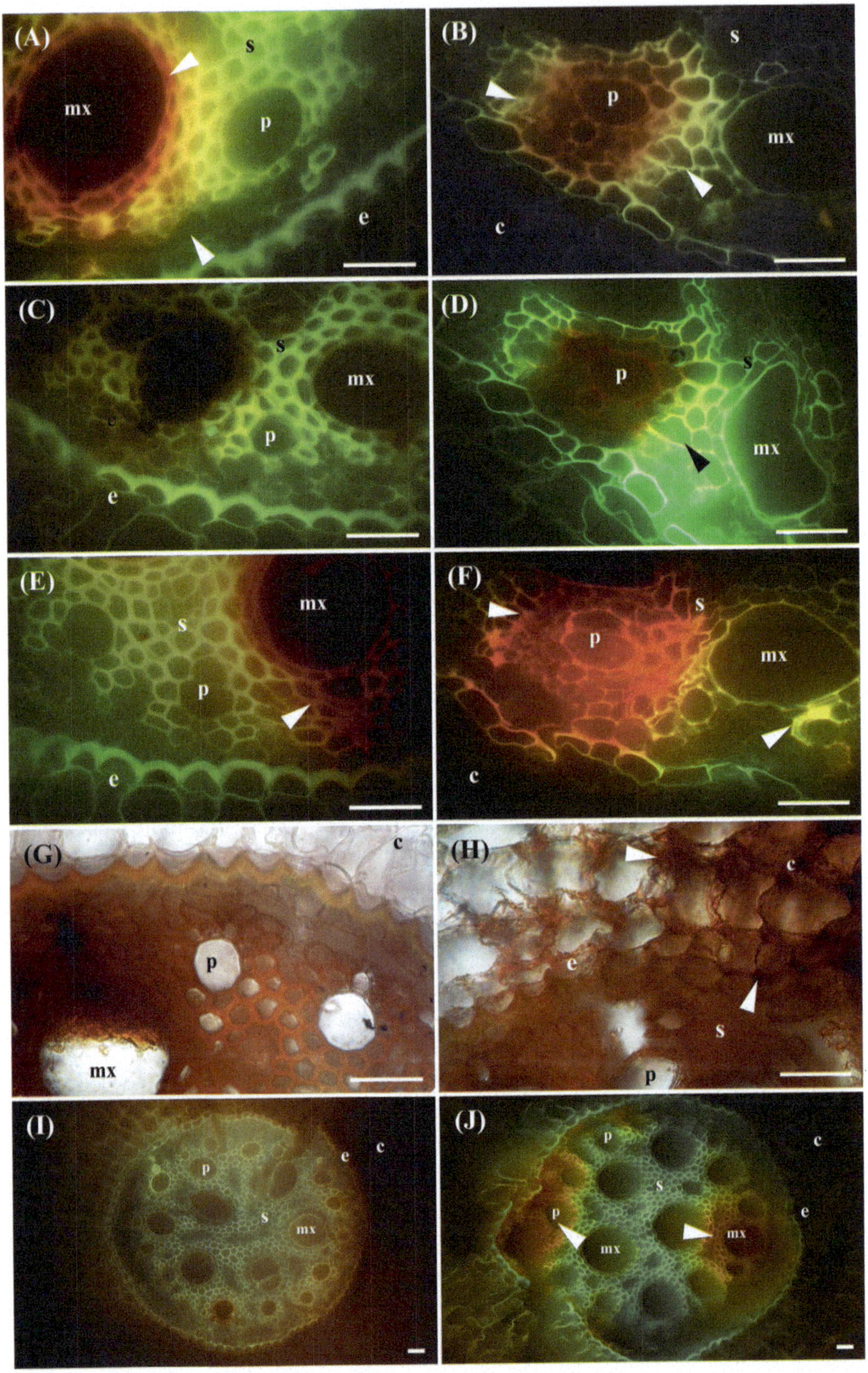

Fig. 5.2 Histochemical characterization of flavonoids, lignin and dopamine in the roots of banana plants cultivar Maçã supplied (+Si) (**b**, **d**, **f**, **h** and **j**) or non-supplied with silicon (–Si) (**a**, **c**, **e**, **g** and **i**) at 24 days after inoculation with *Fusarium oxysporum* f. sp. *cubense*. (**a**) Metaxylem vessels located in the vascular bundles of the roots of the –Si plants exhibit a *yellow-orange* autofluorescence (*arrow*). (**b**) The roots of +Si plants exhibit slight *yellow-orange* autofluorescence (*arrow*)

inoculated plants that were supplied with silicon, and it was confirmed by the intense orange-yellow fluorescence detected at 24 (Fig. 5.2j) and 32 dai (Fig. 5.3j). Da Silva et al. (2015) investigated whether silicon could enhance the production of flavonoids in wheat leaves in response to *P. oryzae* infection at the histochemical level. According to these authors, a high foliar silicon concentration was correlated with reduced fungal growth inside the epidermal cells. A strong fluorescence, which was an indication of the presence of flavonoids, was detected in the leaf cells of plants that received silicon. According to Carver et al. (1998), oat plants deprived of silicon showed increased phenylalanine ammonia-lyase activity and, consequently, there was increased accumulation of phenolic compounds in the epidermal cells colonized by *B. graminis* f. sp. *avenae*. In other words, the activation of the energy-expensive phenylpropanoid pathway likely replaced the fortification of epidermal cell walls by silicon. By contrast, Bélanger et al. (2003) found that the greatest cytochemical difference between wheat plants that were and were not supplied with silicon was the extensive deposition of glycosylated phenolics, as determined by cytochemical labeling, in the cell wall of infected epidermal cells in silicon-supplied plants as well as on the extra-haustorial membrane of *B. graminis* f. sp. *tritici*. The autofluorescence of barley epidermal cells upon *B. gramini* f. sp. *hordei* infection showed a hypersensitive response. This finding suggested that phenolics were present and likely occurred before silicon accumulated to neutralize the dead cell contents while providing the strength and integrity of the surrounding epidermal cells (Koga et al. 1988).

Defense responses have also been reported for dicots such as cucumbers when amended with silicon and then infected by fungal pathogens. Menzies et al. (1991) reported that a great number of leaf cells in cucumber plants that were supplied with silicon and inoculated with *Podosphaera xanthii* showed a rapid accumulation of phenolic-like compounds. Biochemical analyses of leaf extracts from cucumbers that received silicon and were inoculated with *P. xanthii* indicated the presence of flavonoids and phenolic acids that accumulated specifically and strongly in a manner typical of phytoalexins (Fawe et al. 1998). The root cells of cucumbers that were supplied with silicon showed a rapid and more extensive accumulation of electron-dense, phenolic-like material with antifungal activity against the root rot pathogen *Pythium ultimum* (Chérif et al. 1992a). Moreover, Chérif et al. (1994) noted an

Fig. 5.2 (continued) on the phloem and metaxylem vessels. (**c**) No fluorescence was observed in the roots of the−Si plants after staining with Neu's reagent. (**d**) Strong *yellow-orange* fluorescence (*arrow*) observed in the phloem to lemon-yellow fluorescence in the sclerenchyma and metaxylem vessels in the roots of the +Si plants after staining with Neu's reagent. (**e**) Metaxylem vessels on the roots of −Si plants stained with Wilson's reagent exhibited slight orange fluorescence (*arrow*). (**f**) Intense *orange yellow* fluorescence (*arrowheads*) in the cells neighboring the vascular bundles of phloem and metaxylem vessels in the roots of +Si plants stained with Wilson's reagent. (**g**) No evidence of lignin deposition in the roots of −Si plants after staining with phloroglucinol-HCl. (**h**) Strong lignin deposition in the cortex of roots of +Si plants stained with phloroglucinol-HCl (*arrow*). (**i**) Absence of dopamine in the roots of −Si plants. (**j**) Dopamine was strongly suspected to occur in the vascular bundles of phloem and metaxylem vessels (*arrow*) of roots of +Si plants as confirmed by *orange-yellow* fluorescence after staining with lactic acid+glyoxylic acid stain. *c* cortex, *e* endodermis, *mx* metaxylem, *p* phloem, and *s* sclerenchyma. Bars=50 μm (Reproduced from Fortunato et al. (2014), with permission from the American Phytopathological Press)

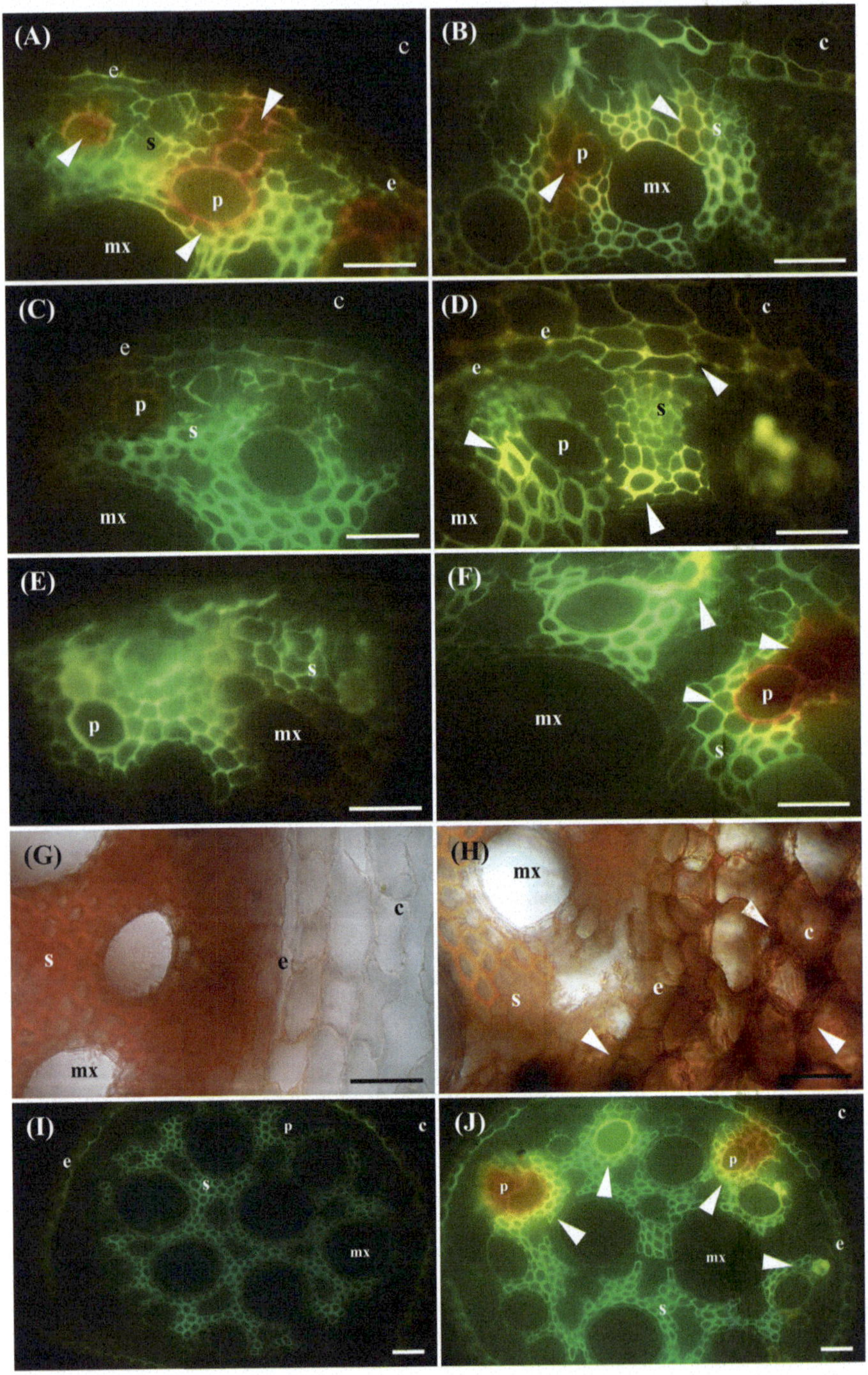

Fig. 5.3 Histochemical characterization of flavonoids, lignin and dopamine in the roots of banana plants cultivar Maçã supplied (+Si) (**b**, **d**, **f**, **h** and **j**) or non-supplied with silicon (–Si) (**a**, **c**, **e**, **g** and **i**) at 32 days after inoculation with *Fusarium oxysporum* f. sp. *cubense*. (**a**) Strong *yellow-orange* autofluorescence (*arrow*) in the vascular bundles and in the sclerenchyma cells in the roots of –Si plants. (**b**) Slight *yellow-orange* autofluorescence (*arrow*) observed near the phloem and metaxylem vessels in the roots of +Si plants. (**c**) Absence of fluorescence in the vascular

increase in the activities of chitinases, peroxidases and polyphenoloxidases in cucumber leaves from plants that received silicon and were inoculated with *P. ultimum* than in infected plants that did not receive silicon. Additionally, leaf extracts from plants that were supplied with silicon and inoculated with *P. ultimum* showed a marked increase in the concentration of antifungal phenolic compounds. Dann and Muir (2002) reported that pea seedlings that received silicon showed an increase in chitinase and β-1,3-glucanase activities prior to *Mycosphaerella pinodes* inoculation. In addition, few lesions were observed on pea leaf seedlings that were supplied with silicon than on the seedlings that were not supplied with this element. In an incompatible cowpea-*Uromyces vignae* interaction, Heath (1981) observed that silicon was associated with electron-opaque regions of the haustorium encasement, the necrotic host cytoplasm and adjacent host cell walls and wall deposits. Although silicon accumulation was not involved in haustorium formation, this element could have been associated with the phenolics found in many disorganized cowpea epidermal cells. Heath and Stumpf (1986) also observed that in cowpea plants supplied with silicon, fungal development apparently ceased while the penetration peg was traversing the haustorial mother cell wall, often before the peg reached the adjacent silicified plant cell wall. However, in plants that were not supplied with silicon, haustorial mother cells for three out of ten infection sites had already formed a haustorium. In the majority of the remaining infection sites, fungal growth appeared to have ceased before the initiation of a visible penetration peg as well as during a stage of development that was observed when the haustorial mother cell and the host cell wall were bridged by an electron-opaque material. The fact that most penetration pegs stopped their development earlier in plants that did not receive silicon than in plants that received silicon supports the previous suggestion that the higher levels of wall-associated phenolic compounds in the former resulted in a faster inhibition of the hydrolytic enzymes that were released by the fungus and were involved in the formation of the penetration peg. These and other ultrastructural observations suggested that the silicified cell walls in silicon-supplied plants may have reduced the interchange of materials between the host and the fungus. As a consequence, the resulting phenolic materials would restrict the flow of materials to the haustorial mother cell that normally prevents its premature senescence, or they would act as a

Fig. 5.3 (continued) bundles,sclerenchyma and endodermis cells in the roots of −Si plants stained with Neu's reagent. (**d**) Strong *yellow* fluorescence in the roots of +Si plants stained with Neu's reagent (*arrow*). (**e**) Slight *orange-yellow* fluorescence in the cells surrounding the phloem and metaxylem vessels and the sclerenchyma cells in the roots of −Si plants stained with Wilson's reagent. (**f**) Strong *orange-yellow* fluorescence in the phloem vessels and in the sclerenchyma cells in the roots of +Si plants stained with Wilson's reagent (*arrow*). (**g**) Absence of lignin deposition in the roots of −Si plants stained with phloroglucinol-HCl. (**h**) Strong lignin deposition in the cortex of the roots of +Si plants stained with phloroglucinol-HCl (*arrow*). (**i**) Absence of dopamine in the roots of −Si plants. (**j**) Dopamine was strongly suspected to occur in the phloem and metaxylem vessels (*arrow*) of the roots of +Si plants as confirmed by *orange-yellow* fluorescence after staining with lactic acid + glyoxylic acid stain. *c* cortex, *e* endodermis, *mx* metaxylem, *p* phloem, and *s* sclerenchyma. Bars = 50 μm (Reproduced from Fortunato et al. (2014), with permission from the American Phytopathological Press)

physical barrier if the penetration peg reaches the host cell wall. This finding is in accordance with previous reports that silicon might form complexes with organic compounds in the epidermal cell walls, consequently increasing their resistance to degradation by the hydrolytic enzymes and non-host-selective toxins released by plant pathogens (Volk et al. 1958; Inanaga et al. 1995a, b). Li and Heath (1990) found an increase in the number of *U. vignae* haustoria, but they found a reduction in the concentration of silicon and in the intensity of the autofluorescence of mesophyll cell walls when injecting abscisic acid and gibberellic acid into bean leaves. Rose plants that were supplied with silicon showed an increase in the concentration of antimicrobial phenolic acids and flavonoids in response to infection by *Podosphaera pannosa* (Shetty et al. 2011). In addition, phenylpropanoid pathway genes such as those coding for phenylalanine ammonia-lyase, cinnamyl alcohol dehydrogenase and chalcone synthase were up-regulated for rose plants that were supplied with silicon (Shetty et al. 2011).

For incompatible and compatible rice-*P. oryzae* interactions, the differential accumulations of glucanase, peroxidase and PR-1 transcripts were associated with limited fungal colonization in the epidermal cells for a blast-susceptible cultivar (Rodrigues et al. 2005). However, the resistant cultivar responded against fungal penetration by developing a hypersensitive response that was associated with a strong induction of PR-1 and peroxidase transcripts regardless of whether silicon was supplied (Rodrigues et al. 2005). Cai et al. (2008) showed that the lower leaf blast disease severity in rice plants that received silicon was linked to higher activities in peroxidases, polyphenoloxidases and phenylalanine ammonia-lyases. Perennial ryegrass plants that were supplied with silicon and were infected by *P. oryzae* exhibited the increased production of several phenolic acids, including chlorogenic acid and flavonoids, greater peroxidase and polyphenoloxidase activities and higher relative expression levels of the genes encoding phenylalanine ammonia-lyase and lipoxygenase compared with the non-silicon-supplied plants (Rahman et al. 2015). For the pathosystems banana-*Fusarium oxysporum* f. sp. *cubense*, coffee-Meloidogyne *exigua*, cotton-*Colletotrichum gossypii* var. *cephalosporioides*, cotton-*Ramularia areola*, rice-*Monographella albescens*, rice-*R. solani*, soybean-*Phakopsora pachyrhizi* and tomato-*Pseudomonas syringae* pv. *tomato*, the activity of phenylalanine ammonia-lyase, the key enzyme in the phenylpropanoid pathway that is responsible for the production of different types of phenolics with antimicrobial properties, increased when plants were supplied with silicon (Silva et al. 2010; Fortunato et al. 2012; Andrade et al. 2013; Cruz et al. 2013; Curvelo et al. 2013b; Guerra et al., 2013a; Tatagiba et al. 2014; Schurt et al. 2014). The greater phenylalanine ammonia-lyase activity was linked to an increase in the concentrations of total soluble phenolics and lignin-thioglycolic acid derivatives in the leaves of banana, coffee and cotton plants that were supplied with silicon and to a decrease in disease development (Silva et al. 2010; Fortunato et al. 2012; Curvelo et al. 2013b; Guerra et al. 2013a, b). Moreover, the beneficial effect of silicon for suppressing infections in the banana-*F. oxysporum* f. sp. *cubense*, cotton-*Colletotrichum gossypii* var. *cephalosporioides*, cotton-*Phakopsora gossypii*, coffee-*M. exigua*, rice-*Bipolaris oryzae,* rice-*M. albescens*, rice-*R. solani*, sorghum-*Colletotrichum sublineolum*, soybean-*Phakopsora pachyrhizi* and tomato-*Pseudomonas syringae* pv. *tomato* pairings was in part explained by an increase in the activities of defense-related enzymes such as peroxidases, polyphenoloxidases, β-1,3-glucanases, and chitinases as well as by an increase in the anthocyanin concentrations for sorghum (Silva et al. 2010; Dallagnol et al. 2011; Fortunato et al. 2012; Tatagiba et al. 2014; Andrade et al. 2013; Cruz et al. 2013; Guerra et al. 2013a, b; Resende et al. 2013; Schurt et al. 2014). Silva et al. (2012) investigated the effects of silicon (0 and 2 mmol) and manganese (0.5, 2.5 and 10 μmol) rates on the activities of peroxidases, polyphenoloxidases and phenylalanine ammonia-lyases in rice that was infected by *B. oryzae*, and they observed that the activities of these three enzymes were not boosted by silicon at any manganese rate. In some rare cases, silicon may not contribute to increased host resistance to disease. Nascimento et al. (2014) examined the response of the soybean cultivars Bossier and Conquista that were or were not supplied with silicon to frogeye leaf spot (*Cercospora sojina*). These authors looked for defense enzyme activities, cell wall-degrading enzymes produced by the fungus (cellulases, xylanases, pectin methyl esterases and polygalacturonase) as well as concentrations of total soluble phenolics and lignin-thioglycolic acid derivatives. According to their findings, the severity of frogeye leaf spot was greater in the Bossier cultivar (susceptible) than in the Conquista cultivar (resistant) as well as in the plants receiving silicon compared with those that did not receive silicon. Except for the concentrations of total soluble phenolics and lignin-thioglycolic acid derivatives, the activities of the defense enzymes and the cell wall-degrading enzymes did not change for non-inoculated plants that were supplied with silicon regardless of the cultivar. The activities of lipoxygenases, phenylalanine ammonia-lyases, chitinases, and polyphenoloxidases as well as the activities of cell wall-degrading enzymes decreased for the inoculated plants that were supplied with silicon and likely compromised their resistance to frogeye leaf spot.

Schurt et al. (2013a) used analytical pyrolysis coupled with gas chromatography and mass spectrometry to investigate possible changes in the chemical composition of lignin in leaf sheaths for the BR-Irga 409 and Labelle rice cultivars that were not and were supplied with silicon and were infected with *R. solani*. Based on the resulting mass spectra, 33 compounds were identified, ten of which were products from the degradation of carbohydrates and 23 of which were derived from lignin. From the lignin derivatives, eight compounds were of the *p*-hydroxyphenyl type, 11 compounds were of the guaiacyl type and four compounds were of the syringyl type. From the leaf sheaths of both cultivars, the concentrations of lignin (*p-hydroxyphenyl*, syringyl (S) and guaiacyl (G)) were approximately 15 %, regardless of whether silicon was present. There was no increase in the S/G ratio except in the leaf sheaths of BR-Irga 409 that were supplied with silicon and infected with *R. solani*. The high silicon concentration in the leaf sheaths of both cultivars, which in turn resulted in an increase in the S/G ratio, most likely contributed to a reduction in the area under the progress curve for sheath blight. In another study, Schurt et al. (2013b) investigated the effect of silicon on the concentrations of soluble and insoluble lignin and sugars in rice leaf sheaths that were infected by *R. solani*. Based on their results, there was no effect from silicon or fungal inoculation on the concentrations of arabinans, galactans, glucans, mannans, sugars, and xylans for the BR-Irga-409 and Labelle cultivars. In addition, no variation was detected in the concentrations of insoluble, soluble and total lignin between the cultivars. The concentrations of total and insoluble lignin were higher for plants that were supplied with silicon regardless of whether they were inoculated. In conclusion, the authors hypothesized that the rice plants that were supplied with silicon were more resistant to sheath blight because of an increase in the lignifications of the leaf sheath tissues and the lower concentration in total sugars.

The silicon effect on the potentiation of host resistance that leads to an increased synthesis of antimicrobial compounds depends on whether this element is supplied via foliar application or via the roots. Foliar-applied silicon can successfully reduce infections of *Podosphaera xanthii* in cucumbers and melons, *Hemileia vastatrix* in coffee, *Bipolaris oryzae* in rice and *Pseudomonas syringae* pv. *tomato* in tomato by affecting pathogen penetration, but this application is never as effective as silicon root applications (Carré-Missio et al. 2009; Rezende et al. 2009; Liang et al. 2005; Andrade et al. 2013; Dallagnol et al. 2015). Because no silicon transporter genes have been identified to date for plant foliage, this foliar effect is likely related to the formation of a physical barrier after the deposition of the material on the leaf surface and/or by an osmotic or pH effect on germinating conidia. Rezende and colleagues (2009) partially demonstrated this finding for brown spot development in rice by showing that foliar applications of silicon to the adaxial leaf surface had practically the same x-ray microanalysis intensity for silicon as applying silicon to the roots. However, for the abaxial leaf surface, the foliar silicon x-ray microanalysis was identical to the non-amended control, whereas applying silicon to the roots expressed the same x-ray microanalysis intensity as the adaxial side of the same treatment. When these authors compared the silicon concentrations in rice tissue among the non-amended control, foliar-applied and applying silicon to the roots treatments, only root-supplied silicon showed significant plant tissue uptake of this element. Furthermore, both Liang et al. (2005) and Dallagnol et al. (2015) demonstrated that when comparing foliar with applying silicon to the roots, only the root applications and not the foliar applications of silicon potentiated plant defense responses such as the activities of peroxidases, polyphenoloxidases, β-1,3-glucanases and chitinases (Pereira et al. 2009a, b; Liang et al. 2005; Carré-Missio et al. 2009; Andrade et al. 2013; Dallagnol et al. 2015). Proposed models for the modes of action of potassium silicate, a source of soluble silicon, is shown in Fig. 5.4a and b for foliar or root applications.

Fig. 5.4 (continued) region without silica bodies. The host defense is potentiated by the presence of soluble silicon at the fungal infection site. *Continuous blue lines* indicate the stimulated route; *dashed green lines* indicate the probable stimulated route; *continuous red lines* indicate the repressed route; *dashed red lines* indicate the probable repressed route; *gray lines* indicate routes that are not directly affected. *ap* appressorium, *CAT* catalase, *CHI* chitinase, *cn* conidium, *ct* cuticle, *cw* cell wall, *da* direct action on pathogen, *GLU* β-1,3-glucanases, H_2O_2 hydrogen peroxide, *lp* lipid peroxidation, O_2^- superoxide anion, OH^- hydroxyl radical, *ph* penetration hyphae, *pm* plasma membrane, *POX* peroxidases, *PPO* polyphenoloxidases, *Psp* polymerized potassium silicate, *Si* soluble silicon, *SOD* superoxide dismutases and *tm* transition metal

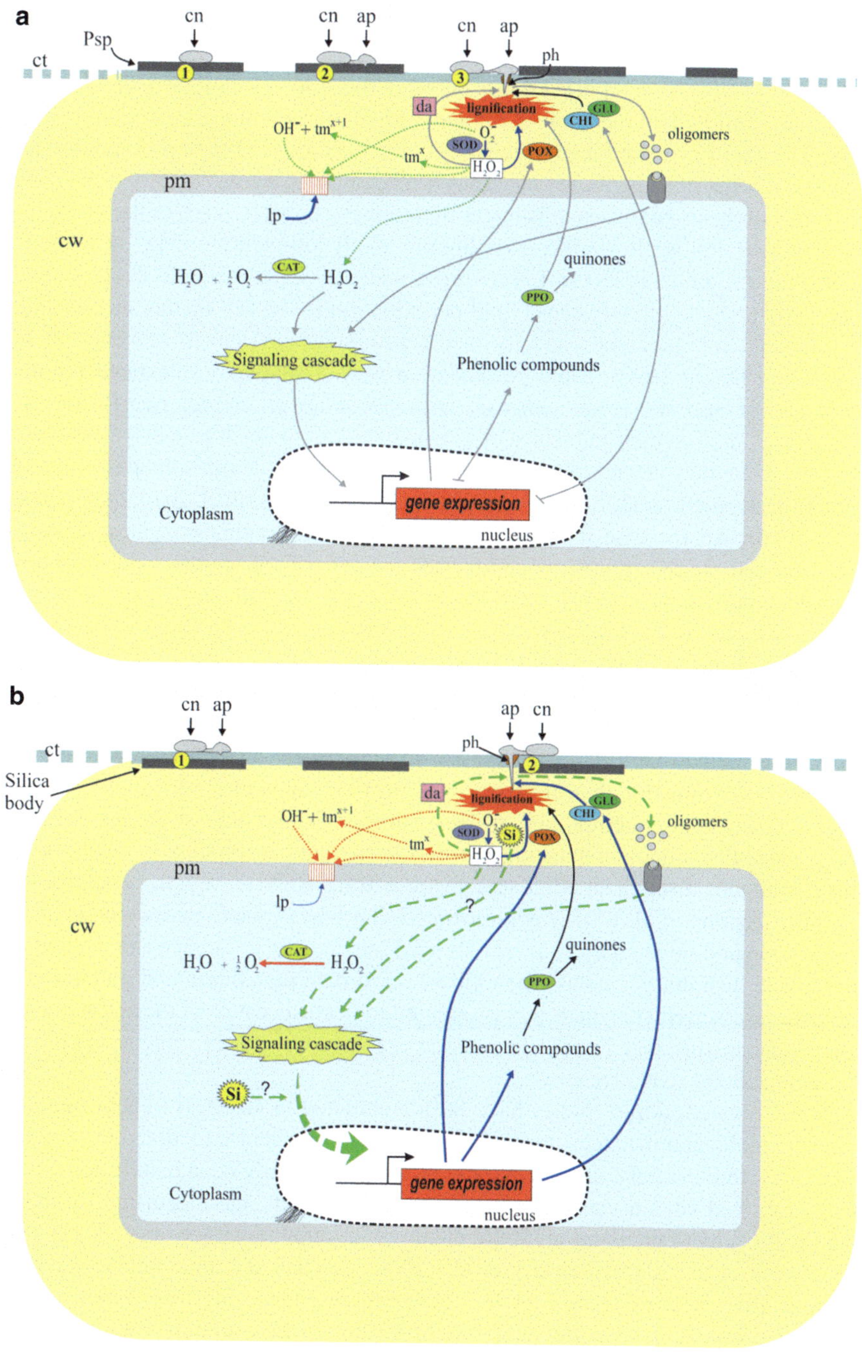

Fig. 5.4 Models proposed for the modes of action of potassium silicate (PS) when applied to the leaves (**a**) or via the roots (**b**). The numbers 1, 2 and 3 on top of the *yellow circle* represent the host defense mechanisms. (**a**) *1* – conidial germination is inhibited due to the deposition of PS on the leaf surface, *2* – the penetration peg is inhibited due to the deposition of PS on the leaf surface, *3* – a leaf region without the deposition of PS favor fungal penetration; (**b**) *1* – the fungal peg penetration is inhibited due to the presence of silica bodies inside the epidermal cells, *2* – a leaf

Several studies have reported a link between the silicon supply and an improvement in the antioxidant metabolism of plants when they are infected by plant pathogens. The rapid production of reactive oxygen species (ROS) in the apoplast in response to infections by these pathogens has been proposed as one way in which a plant may orchestrate the establishment of defensive barriers, such as the strengthening of host cell walls via the cross-linking of glycoprotein, to delay host tissue colonization (Lamb and Dixon 1997; Torres et al. 2006). However, ROS are known to be toxic and can directly cause lipid peroxidation in the cell membrane, leading to a demand for increased capacity in the antioxidant system to scavenge them (Lamb and Dixon 1997). Lipid peroxidation was dramatically alleviated for the banana-*F. oxysporum* f. sp. *cubense*, cotton-*Ramularia areola*, rice-*P. oryzae*, sorghum-*C. sublineolum* and wheat-*P. oryzae* interactions, as indicated by the lower malonic aldehyde concentration in plants that were supplied with silicon (Fortunato et al. 2012; Resende et al. 2012; Curvelo et al. 2013a; Debona et al. 2014; Domiciano et al. 2015). An increase in the activities of ROS-scavenging enzymes, such as ascorbate peroxidases, glutathione reductases, superoxide dismutases and catalases in plants receiving silicon restricted the ROS-dependent cellular damage that was indirectly linked to the high concentration of malonic aldehyde (Mohaghegh et al. 2011; Sun et al. 2010; Li et al. 2012; Resende et al. 2012; Curvelo et al. 2013a; Polanco et al. 2014; Domiciano et al. 2015).

In a proteomic analysis, Liu et al. (2014) found that the quantities of ascorbate peroxidase, dehydroascorbate reductase and superoxide dismutase were reduced after *P. oryzae* infection, but they increased for rice plants that were supplied with silicon. Collectively, the findings of these authors clearly suggest the pivotal role that is played by silicon in managing the ROS generated in response to infection by plant pathogens through an efficient activation of the ROS-scavenging systems. By contrast, Debona et al. (2014) demonstrated that wheat plants that were supplied with silicon and infected by *P. oryzae* showed lower cellular damage and decreased superoxide dismutase, catalase, peroxidase, ascorbate peroxidase and glutathione-S-transferase activities, which was postulated to occur because of the activation of other mechanisms that limited leaf tissue colonization by the fungus, therefore reducing cellular oxidative stress.

A few transcriptomic studies have been conducted in an effort to better understand the molecular mechanisms of silicon-mediated resistance to infection by plant pathogens. Interestingly, the beneficial effect has so far only been manifested when plants received silicon via the roots and were then challenge-inoculated by these plant pathogens (Fauteux et al. 2006; Chain et al. 2009; Ghareeb et al. 2011; Van Bockhaven et al. 2015). Fauteux et al. (2006) conducted a microarray study to examine the effect of silicon on the increased resistance of *Arabidopsis* plants against *B. graminis* f. sp. *tritici* infection. According to these authors, the expression of all but two genes was unaffected by silicon for non-inoculated plants. By contrast, for inoculated plants that were not and were supplied with silicon, the expression of a set of nearly 4,000 genes was dramatically altered. After a functional categorization, many of the up-regulated genes were found to be defense-related, whereas a large proportion of down-regulated genes were involved in primary plant metabolism. The regulated defense genes included R genes, stress-related transcription factors, genes involved in signal transduction, the biosynthesis of stress hormones (salicylic acid, jasmonic acid and ethylene) and the metabolism of ROS. In inoculated plants that were supplied with silicon, the magnitude of down-regulation was attenuated by more than 25 % (Fauteux et al. 2006). Chain et al. (2009) performed a large transcriptomic analysis (55,000 unigenes) to compare the differential responses of wheat plants that were not and were supplied with silicon and were not or were inoculated with *B. graminis* f. sp. *tritici*. The response to the silicon supply in the non-inoculated plants was limited to 47 genes with diverse functions, and there was little evidence of the regulation of a specific metabolic process. For the inoculated plants, there was an up-regulation of many genes that were linked to stress and metabolic processes and a down-regulation of genes linked to photosynthesis. For plants that were supplied with silicon and infected by *B. graminis* f. sp. *tritici*, the disease symptoms were reduced and translated into a nearly perfect reversal of genes that were regulated in infected plants that had not received silicon. According to these authors, silicon played a limited role in the wheat transcriptome in the absence of fungal infection. However, the benefits of silicon in reducing disease symptoms were remarkably aligned with a counter-response to transcriptomic changes upon fungal infection. According to the microarray analysis of tomato plants that were infected with *Ralstonia solanacearum* as conducted by Ghareeb et al. (2011), there was an up-regulation in the expression of jasmonic acid/ethylene marker genes, oxidative stress marker genes and basal defense marker in the presence of silicon. These findings help to explain partly why the wilt symptoms caused by *R. solanacearum* were dramatically reduced.

Brunings and her colleagues (2009) described the effect of silicon on the molecular response of rice to *P. oryzae* infection on a genome-wide scale when using a 44 k rice microarray to compare gene expression levels between rice plants that were not or were supplied with silicon and were not or were inoculated with *P. oryzae*. The primary purpose of their study was to investigate the interaction between silicon and *P. oryzae* inoculation on the transcriptional profile of rice. They found that defense/stress-related genes were differentially up and down-regulated in *P. oryzae* comparisons with silicon + *P. oryzae* treatments. These comparisons were of particular interest because they highlighted how silicon changed plant reactions to fungal infections. Among these defense/stress-related genes were ethylene signaling pathway genes, a gene encoding a thaumatin/pathogenesis-related protein (Os12 g0568900), a class III peroxidase (Os07 g0677500) and a number of transcription factors and protein kinases. The authors further noted that in addition to simply attenuating the plant response to *P. oryzae* infection, a substantially different pattern of expression was noted in their experiment. Not only did the silicon comparison with the silicon + *P. oryzae* treatment reveal 440 differentially expressed genes less than the control comparison with the *P. oryzae* treatment, but the two comparisons had only 236 genes in common. Silicon therefore affected the interactions between the host and the pathogen at the molecular level by attenuating the rice response to the pathogen and by influencing the differential expression of a unique set of genes. Silicon was clearly responsible for preconditioning plants to react to stress, which was supported in this study by the fact that *P. oryzae* infection resulted in less than half the number of differentially expressed genes in plants that were supplied with silicon than those that were not supplied with silicon (298 compared with 738).

Pursuant to the fact that photorespiration is important in plants that are under biotic stress, diatom photorespiration has been shown to be dependent on silicon polymerization, and the activities of photorespiratory enzymes were higher in plants supplied with silicon under stress conditions. Von Bonkhaven et al. (2013) proposed a hypothesis in which photorespiration might play an important role in rice resistance to brown spot as caused by *B. oryzae* in the presence of silicon. A transcriptome study conducted by Van Bockhaven et al. (2015) showed that fungal infections repressed photosynthesis and lowered nitrate concentrations in plants that were not supplied with silicon, which resulted in greater brown spot symptom development. By contrast, for plants that were supplied with silicon, there was an up-regulation of several photorespiratory marker genes, leading to the hypothesis that increased photorespiration rates may be one of the driving forces behind the possible effects of silicon on rice photosynthesis (Van Bockhaven et al. 2015).

It has been suggested that silicon primes the plant immune response rather than constitutively inducing defense-related genes that result in an increase in host resistance (an increase in the magnitude of host defensive processes and/or the speed with which they are activated) after the plant has been challenged by a plant pathogen (Ghareeb et al. 2011; Van Bockhaven et al. 2013, 2015; Dallagnol et al. 2015; Vivancos et al. 2015). This primed state allows the plant to respond more quickly and effectively to challenges because of the accumulation of inactive cellular proteins that are involved in signal transduction such as mitogen-activated protein kinases, chromatin modifications and alterations in primary metabolism with a minimal metabolic cost (Van Hulten et al. 2006; Conrath 2011). According to Van Bockhaven et al. (2013), the broad-spectrum disease resistance found in rice plants that received silicon was related, at least in part, to the priming effect that resulted in a differential accumulation of defense-regulatory transcription factors, a process that was sufficient for priming defense genes but less effective at activating them directly. An additional mechanism underpinning the potentiation of host resistance to pathogen infection by silicon has been the involvement of this element in plant hormone responses. This response has been observed for host-pathogen interactions that are mediated by salicylic acid, jasmonate and ethylene in the *Arabidopsis-Erysiphe cichoracearum* interaction (Fauteux et al. 2006), the ethylene in rice-*P. oryzae* interaction (Brunings et al. 2009), the jasmonate and ethylene in tomato-*R. solanacearum* interaction (Ghareeb et al. 2011) and the ethylene in rice-*B. oryzae* interaction (Van Bockhaven et al. 2015). For the rice-*B. oryzae* interaction, the increased resistance of plants that were supplied with silicon to brown spot was attributed to the production and/or action of fungal ethylene that impaired the fungus' ability to suppress the rice innate immune system and, as a consequence, resulted in a faster and stronger activation of the basal mechanisms of host defense

(Van Bockhaven et al. 2015). Vivancos et al. (2015) engineered *Arabidopsis* plants with a silicon transporter from wheat (TaLsi1) and exploited mutants (*pad4* and *sid2*) that were deficient in salicylic acid-dependent defense responses. The purpose of this transporter engineering and mutant exploitation was to study the phenotypic response and changes in defense expressions against *Golovinomyces cichoracearum* infection when plants were amended with silicon. According to these authors, the *TaLsi1* plants exhibited significantly greater concentrations of silicon in plant tissue and were significantly more resistant to infection by *G. cichoracearum* than the non-inoculated control plants that were supplied with silicon. The resistant plants accumulated higher levels of salicylic acid and expressed higher levels of transcripts encoding for defense-related genes. However, *TaLsi1 pad4* and *TaLsi1 sid2* plants were also more resistant to *G. cichoracearum* infection than *pad4* and *sid2* plants in the presence of silicon. An analysis of the resistant phenotypes revealed a significant reduction in the production of salicylic acid and the expression of defense genes in comparison with those of the susceptible controls. The results obtained by these authors indicated that silicon contributed to *Arabidopsis* defense priming following *G. cichoracearum* infection, and they further highlighted that silicon could confer protection even when the priming was altered.

Silicon-Mediated Host Resistance to Pathogens Through Changes in the Primary Metabolism

Changes in the growth and development of plants are the results of the occurrence of and constant exposure to several abiotic and biotic stresses (Berger et al. 2007). Biotic stresses, particularly those caused by plant pathogens, lead to changes in primary plant metabolism. This response will in turn provide energy for the host's defense responses that originate from secondary metabolism and are primarily based on activating the expression of hundreds of genes that are involved in many defense pathways (Berger et al. 2007; Rojas et al. 2014). During pathogen infection, the host's physiology is negatively affected primarily because of changes in leaf gas exchange, once the amount of healthy leaf area is decreased and the efficiency of the photosynthetic process is dramatically lowered (Shtienberg 1992). The most notable negative effects that have resulted from pathogen infections of their hosts are the reduced concentration of pigments, damage to the chloroplasts, impairments in energy dissipation via chlorophyll *a* fluorescence and an increase in the foliar temperature (Petit et al. 2006; Zhao et al. 2011). For instance, genes involved in photosynthesis and chlorophyll biosynthesis have been found to be down-regulated in many different host-parasite interactions (Scholes and Rolfe 1996; Berger et al. 2004; Swarbrick et al. 2006; Bilgin et al. 2010). By contrast, several genes that are involved in energy production, such as glycolysis and the pentose phosphate pathways, the tricarboxylic acid cycle, mitochondrial electron transport and ATP biosynthesis become up-regulated during the infection time-course by a plant pathogen (Less et al. 2011; Rojas et al. 2014).

Although the aforementioned examples are now well-known, our current understanding of how silicon affects the physiology and biochemistry of plants that are infected by plant pathogens remains to be elucidated. For this reason, research efforts over the last few years have focused on examining the role played by this element in host physiology and primary metabolic pathways, especially for alterations in photosynthesis that occur during the infection process of several plant pathogens in crops of economic importance. For sorghum-*Colletotrichum sublineolum* and common bean-*C. lindemuthianum* interactions, in addition to the reduction of anthracnose symptoms in plants supplied with silicon, the values for the net carbon assimilation rate (*A*), stomatal conductance to water vapor (g_s) and transpiration rate (*E*) were higher for infected plants that had received silicon than for infected plants that were not supplied with this element. These findings suggested that the physiology of sorghum and common bean plants was negatively impaired upon pathogen infection, but it was greatly reduced in the presence of silicon. There were no changes in the *A*, g_s and *E* for the non-infected plants supplied with silicon. Furthermore, the impaired photosynthetic performance of plants that received silicon was deeply associated with stomatal limitations, whereas in the non-infected plants, those impairments likely reflected dysfunctions at the biochemical reaction level that were involved in CO_2 fixation (Resende et al. 2012; Polanco et al. 2014). Under field conditions, Rodrigues et al. (2015) reported that no difference was detected between bean plants that were not and were sprayed with potassium silicate (KSi) with respect to the *A*, g_s, *E* and internal CO_2 concentration. However, the *A* significantly increased by 17 % in plants that were treated with the fungicide azoxystrobin. The *A* was not affected by KSi or sodium molybdate (NaMo); however, the *A* was significantly increased by 13 % after NaMo + KSi applications. In conclusion, bean plants that were sprayed with KSi and NaMo were associated with decreased anthracnose severity as well as enhanced photosynthesis.

For the rice- and wheat-*P. oryzae* interactions, higher *A*, g_s and *E* values were obtained for infected plants that were supplied with silicon in contrast to the lower values of infected plants that were not supplied with this element (Aucique-Perez et al. 2014; Rios et al. 2014; Domiciano et al. 2015). Biochemical dysfunctions at the chloroplast level likely played a key role in limiting *A* upon *P. oryzae* infection for both rice and wheat plants instead of causing diffusive (stomatal) limitations to photosynthesis. Higher *A*, g_s and *E* values and an increased concentration of leaf pigments were reported to occur in cotton plants that were supplied with silicon and infected by *Ramularia areola* (Curvelo et al. 2013a) and *Colletotrichum gossypii* var. *cephalosporioides* (Guerra et al. 2013a).

By measuring the emission of chlorophyll *a* fluorescence, which is considered to be a powerful tool and a very sensitive probe for the physiological status of plants (Baker 2008), the authors demonstrated that some photochemical parameters such as the quantum yield of photosystem II (PSII) photochemistry (F_v/F_m), photochemical quenching coefficient (q_p) and electron transport rate (ETR), together with the quantification of chlorophylls and carotenoid concentrations, were greatly improved for rice and wheat plants that were supplied with silicon. By contrast, the heat dissipation of the chlorophyll excitation energy, which is measured on the basis of the non-photochemical quenching (NPQ) parameter, decreased for rice and wheat plants that were supplied with silicon and inoculated with *P. oryzae*. Therefore, the PSII electron transport at the chloroplast level was not impaired and the photoprotective processes were kept at the desired physiological level, thus contributing to the integrity of the photosynthetic apparatus. According to Gao et al. (2011), an increase in rice resistance to *P. oryzae* infection from silicon was associated with an enhancement in the photochemical efficiency, more specifically, an increase in the maximum/potential quantum efficiency (F_v/F_m) and the maximum primary yield (F_v/F_0) of the photochemistry of PSII.

Microarray and proteome techniques have been exploited in an attempt to increase our current knowledge regarding the beneficial effects of silicon on the physiology of several plant species during pathogen infection at the molecular level. A transcriptome analysis of silicon's effect on powdery mildew (*Erysiphe cichoracearum*) development in *Arabidopsis thaliana* plants indicated that several genes that were involved in primary metabolism such as photosynthesis and energy pathways as well as amino acid, carbohydrate and lipid metabolism were down-regulated as a direct result of fungal infection. However, many of these same genes, particularly those involved in photosynthesis and energy pathways, were less responsive for plants that were supplied with silicon (Fauteux et al. 2006). Chain et al. (2009) performed a transcriptomic analysis of the silicon effect on a *B. graminis* f. sp. *tritici* infection for wheat and found that many genes that were associated with stress and metabolic processes were up-regulated for infected plants, and genes related to photosynthesis were down-regulated. Conversely, when plants were supplied with silicon prior to fungal inoculation, the genes that were associated with stress and metabolic processes were down-regulated and the genes linked to photosynthesis were up-regulated. In conclusion, the authors noted that the stress imposed by fungal infection was greatly diminished in the presence of silicon. A proteomic study performed by Liu et al. (2014) to examine the effect of silicon on rice resistance to *P. oryzae* infection indicated that the pattern of protein spots was greatly affected by fungal infection regardless of the presence of silicon. Many proteins related to photosynthesis (the chlorophyll *a*/*b*-binding protein, chloroplast putative thylakoid lumenal 16.5 kDa protein, sedoheptulose-1,7-bisphosphatase and ribulose bisphosphate carboxylase large chain) were down-regulated upon fungal infection. In the presence of silicon and fungal infection, these proteins were all up-regulated. These photosynthesis-related proteins in silicon-mediated higher abundance as mediated by silicon may function as light receptors or they may play a role in protein biosynthesis at the chloroplast level, thus affecting rice photosynthesis. Moreover, the differential expression of energy metabolism-related proteins that are involved in the tricarboxylic acid or the pentose phosphate pathways may increase rice resistance against *P. oryzae* infection.

Because the activation of host defense responses requires a large amount of energy together with the induction of the sink metabolism, the photosynthetic capacity and carbohydrate metabolism can be negatively impacted in response to pathogen infection (Ehness et al. 1997). Dallagnol et al. (2013) investigated the effect of silicon uptake on the photosynthesis and leaf sugar concentration in rice plants from the Oochikara cultivar and the *lsi1* mutant (*l*ow-*si*licon; defective in the active silicon uptake) that were not or were infected with *B. oryzae*. The inefficiency of the *lsi1* mutant plants in actively taking up silicon negatively affected rice resistance against *B. oryzae* infection, and it reduced photosynthesis and the sugar concentration. However, the high foliar silicon concentration resulted in an increase in the soluble sugar concentration, photosynthesis, and, consequently, rice resistance to *B. oryzae* infection. The authors concluded that a minimum silicon concentration was needed in the leaf tissue of rice plants to avoid the negative impacts of *B. oryzae* infection on photosynthesis and the sugar concentration. Indeed, rice resistance to brown spot was independently and additively affected by the silicon and soluble sugar concentrations in the leaf tissue (Dallagnol et al. 2013).

Conclusions

In spite of recent advances linking silicon to host resistance via the -omics, i.e., genomics, proteomics and metabolomics, the exact mechanism(s) by which this element modulates plant physiology through an increase in host resistance still requires further investigation. The information generated to date has provided novel insights into silicon's potential to interact with multiple pathways in the plant's primary metabolism to cope better with infections caused by both soil-borne and foliar pathogens. In considering the current plant nutriomics scenario, it remains to be determined as to whether the involvement of silicon in plant-signaling pathways leads to the potentiation of host defense mechanisms and simultaneously makes it feasible to modify some key regulator(s) to enhance silicon uptake. In the near future, the real functions of silicon will be possible to elucidate at the molecular, cellular, organ and even whole plant levels. The function of silicon as linked through enzyme antioxidants and photosynthesis would be a few of the targeted focus areas, and thus the use of this quasi-essential element may be enhanced for sustainable plant health and plant performance.

References

Abed-Ashtiani F, Kadir JB, Selamat AB et al (2012) Effect of foliar and root application of silicon against rice blast fungus in MR219 rice variety. Plant Pathol J 28:164–171. doi:10.5423/PPJ.2012.28.2.164

Andrade CCL, Resende RS, Rodrigues FA et al (2013) Silicon reduces bacterial speck development on tomato leaves. Trop Plant Pathol 38:436–442. doi:10.1590/S1982-56762013005000021

Aucique-Perez CEA, Rodrigues FA, Moreira WR, DaMatta FM et al (2014) Leaf gas exchange and chlorophyll *a* fluorescence in wheat plants supplied with silicon and infected with *Pyricularia oryzae*. Phytopathology 104:143–149. doi:10.1094/PHYTO-06-13-0163-R

Baker NR (2008) Chlorophyll fluorescence: a probe of photosynthesis *in vivo*. Annu Rev Plant Biol 59:89–113. doi:10.1146/annurev.arplant.59.032607.092759

Bélanger RR, Benhamou N, Menzies JG (2003) Cytological evidence of an active role of silicon in wheat resistance to powdery mildew (*Blumeria graminis* f. sp. *tritici*). Phytopathology 93:402–412. doi:10.1094/PHYTO.2003.93.4.402

Berger S, Papadopoulos M, Schreiber U et al (2004) Complex regulation of gene expression, photosynthesis and sugar levels by pathogen infection in tomato. Physiol Plant 122:419–428. doi:10.1111/j.1399-3054.2004.00433.x

Berger S, Sinha AK, Roitsch T (2007) Plant physiology meets phytopathology: plant primary metabolism and plant-pathogen interactions. J Exp Bot 58:4019–4026. doi:10.1093/jxb/erm298

Bilgin DD, Zavala JA, Zhu J et al (2010) Biotic stress globally downregulates photosynthesis genes. Plant Cell Environ 33:1597–1613. doi:10.1111/j.1365-3040.2010.02167.x

Brunings AM, Datnoff LE, Ma JF et al (2009) Differential gene expression of rice in response to silicon and rice blast fungus *Magnaporthe oryzae*. Ann Appl Biol 155:161–170. doi:10.1111/j.1744-7348.2009.00347.x

Cacique IS, Domiciano GP, Moreira WR et al (2013) Effect of root and leaf applications of soluble silicon on blast development in rice. Bragantia 72:304–309. doi:10.1590/brag.2013.032

Cai K, Gao D, Luo S et al (2008) Physiological and cytological mechanisms of silicon-induced resistance in rice against blast disease. Physiol Plant 134:324–333. doi:10.1111/j.1399-3054.2008.01140.x

Carré-Missio V, Rodrigues FA, Schurt DA et al (2009) Ineficiência do silício no controle da ferrugem do cafeeiro em solução nutritiva. Trop Plant Pathol 34:416–421. doi:10.1590/s1982-56762009000600009

Carver TLW, Zeyen RJ, Ahlstrand GG (1987) The relationship between insoluble silicon and success or failure of attempted primary penetration by powdery mildew (*Erysiphe graminis*) germlings on barley. Physiol Mol Plant Pathol 31:133–148. doi:10.1016/0885-5765(87)90012-9

Carver TLW, Robbins MP, Thomas BJ et al (1998) Silicon deprivation enhances local autofluorescent responses and phenylalanine ammonia lyase activity in oat attacked by *Blumeria graminis*. Physiol Mol Plant Pathol 52:245–257. doi:10.1006/pmpp.1998.0149

Chain F, Côté-Beaulieu C, Belzile F et al (2009) A comprehensive transcriptomic analysis of the effect of silicon on wheat plants under control and pathogen stress conditions. Mol Plant Microbe Interact 22:1323–1330. doi:10.1094/MPMI-22-11-1323

Chérif M, Benhamou N, Menzies JG et al (1992a) Silicon induced resistance in cucumber plants against *Pythium ultimum*. Physiol Mol Plant Pathol 41:411–425. doi:10.1016/0885-5765(92)90053-X

Chérif M, Menzies JG, Benhamou N, Bélanger RR et al (1992b) Studies of silicon distribution in wounded and *Pythium ultimum* infected cucumber plants. Physiol Mol Plant Pathol 41:371–385. doi:10.1016/0885-5765(92)90022-N

Chérif M, Asselin A, Bélanger RR (1994) Defense responses induced by soluble silicon in cucumber roots infected by *Pythium* spp. Phytopathology 84:236–242. doi:10.1094/Phyto-84-236

Conrath U (2011) Molecular aspects of defence priming. Trends Plant Sci 16:524–531. doi:10.1016/j.tplants.2011.06.004

Cruz MFA, Rodrigues FA, Polanco LR et al (2013) Inducers of resistance and silicon on the activity of defense enzymes in the soybean-*Phakopsora pachyrhizi* interaction. Bragantia 72:162–172. doi:10.1016/1369-5266(88)80053-0

Curvelo CRS, Rodrigues FA, Pereira LF et al (2013a) Trocas gasosas e estresse oxidativo em plantas de algodoeiro supridas com silício e infectadas por *Ramularia areola*. Bragantia 72:346–359. doi:10.1590/brag.2013.053

Curvelo CRS, Rodrigues FA, Silva LC et al (2013b) Mecanismos bioquímicos da defesa do algodoeiro à mancha de ramularia mediados pelo silício. Bragantia 72:41–51

Da Silva WL, Cruz MFA, Fortunato AA, Rodrigues FA (2015) Histochemical aspects of wheat resistance to leaf blast mediated by silicon. Sci Agric 72(4):322–327

Dallagnol LJ, Rodrigues FA, DaMatta FM et al (2011) Deficiency in silicon uptake affects cytological, physiological, and biochemical events in the rice-*Bipolaris oryzae* interaction. Phytopathology 101:92–104. doi:10.1094/PHYTO-04-10-0105

Dallagnol LJ, Rodrigues FA, Chaves ARM, DaMatta FM (2013) Photosynthesis and sugar concentration are impaired by the defective active silicon uptake in rice plants infected with *Bipolaris oryzae*. Plant Pathol 62:120–129. doi:10.1111/j.1365-3059.2012.02606.x

Dallagnol LJ, Rodrigues FA, Pascholati SF et al (2015) Comparison of root versus foliar applied potassium silicate in potentiating postinfection defences of melon against powdery mildew. Plant Pathol. doi:10.1094/PHYTO.2003.93.4.402

Dann EK, Muir S (2002) Peas grown in media with elevated plant-available silicon levels have higher activities of chitinase and β-1,3-glucanase, are less susceptible to a fungal leaf spot pathogen and accumulate more foliar silicon. Australas Plant Pathol 31:9–13

Datnoff LE, Rodrigues FA (2005) The role of silicon in suppressing rice diseases. APSnet Features. doi:10.1094/APSnetFeature-2005-0205

Debona D, Rodrigues FA, Rios JA et al (2014) The effect of silicon on antioxidant metabolism of wheat leaves infected by *Pyricularia oryzae*. Plant Pathol 63:581–589. doi:10.1111/ppa.12119

Domiciano GP, Rodrigues FA, Guerra AMN et al (2013) Infection process of *Bipolaris sorokiniana* on wheat leaves is affected by silicon. Trop Plant Pathol 38:258–263. doi:10.1590/S1982-56762013005000006

Domiciano GP, Cacique IS, Freitas CC et al (2015) Alterations in gas exchange and oxidative metabolism in rice leaves infected by *Pyricularia oryzae* are attenuated by silicon. Phytopathology 105:738–747. doi:10.1094/PHYTO-10-14-0280-R

Ehness R, Ecker M, Godt DE et al (1997) Glucose and stress independently regulate source and sink metabolism and defense mechanisms via signal transduction pathways involving protein phosphorylation. The Plant Cell Online 9:1825–1841. doi:10.1105/tpc.9.10.1825

Fauteux F, Rémus-Borel W, Menzies JG et al (2005) Silicon and plant disease resistance against pathogenic fungi. FEMS Microb Lett 249:1–6. doi:10.1016/j.femsle.2005.06.034

Fauteux F, Chain F, Belzile F et al (2006) The protective role of silicon in the Arabidopsis–powdery mildew pathosystem. Proc Natl Acad Sci U S A 103:17554–17559. doi:10.1073/pnas.0606330103

Fawe A, Abou-Zaid M, Menzies JG, Bélanger RR et al (1998) Silicon-mediated accumulation of flavonoid phytoalexins in cucumber. Phytopathology 88:396–401. doi:10.1094/PHYTO.1998.88.5.396

Fortunato AA, Rodrigues FA, Nascimento KJT (2012) Physiological and biochemical aspects of the resistance of banana plants to Fusarium wilt potentiated by silicon. Phytopathology 102:957–966. doi:10.1094/PHYTO-02-12-0037-R

Fortunato AA, da Silva WL, Rodrigues FA (2014) Phenylpropanoid pathway is potentiated by silicon in the roots of banana plants during the infection process of *Fusarium oxysporum* f. sp. *cubense*. Phytopathology 104:597–603. doi:10.1094/PHYTO-07-13-0203-R

Freeman BC, Beattie GA (2008) An overview of plant defenses against pathogens and herbivores. Plant Health Instruc. doi:10.1094/PHI-I-2008-0226-01

Gao D, Cai K, Chen J et al (2011) Silicon enhances photochemical efficiency and adjusts mineral nutrient absorption in *Magnaporthe oryzae* infected rice plants. Acta Physiol Plant 33:675–682. doi:10.1007/s11738-010-0588-5

Ghareeb H, Bozsó Z, Ott PG et al (2011) Transcriptome of silicon-induced resistance against *Ralstonia solanacearum* in the silicon non-accumulator tomato implicates priming effect. Physiol Mol Plant Pathol 75:83–89. doi:10.1016/j.pmpp.2010.11.004

Guerra AMNM, Rodrigues FA, Berger PG et al (2013a) Aspectos bioquímicos da resistência do algodoeiro à ramulose potencializada pelo silício. Bragantia 72:292–303. doi:10.1590/brag.2013.036

Guerra AMNM, Rodrigues FA, Berger PG et al (2013b) Resistência do algodoeiro à ferrugem tropical potencializada pelo silício. Bragantia 72:279–291. doi:10.1590/brag.2013.037

Hashioka Y (1942) Studies on rice blast disease in the tropics. (I) Anatomical comparison of leaf epidermis of Formosan rice with that of Taiwan rice from plant pathological viewpoints. Agric Hortic 17:846–852

Hayasaka T, Fujii H, Ishiguro K (2008) The role of silicon in preventing appressorial penetration by the rice blast fungus. Phytopathology 98:1038–1044. doi:10.1094/PHYTO-98-9-1038

He C, Ma J, Wang L (2015) A hemicellulose-bound form of silicon with potential to improve the mechanical properties and regeneration of the cell wall of rice. New Phytol 206:1051–1062

Heath MC (1981) Insoluble silicon in necrotic cowpea cells following infection with an incompatible isolate of the cowpea rust fungus. Physiol Plant Pathol 19:273–276

Heath MC, Stumpf MA (1986) Ultrastructural observations of penetration sites of the cowpea rust fungus in untreated and silicon-depleted French bean cells. Physiol Mol Plant Pathol 29:27–39

Hemmi T, Abe T, Inoue Y (1941) Studies on the rice blast disease: relation of the environment to the development of blast disease and races of the blast fungus. Noji Kairyo Shiryo 157:1–232

Inanaga S, Okasaka A, Tanaka S (1995a) Calcium and silicon binding compounds in cell walls of rice shoots. Jpn J Soil Sci Plant Nutr 41:103–110

Inanaga S, Okasaka A, Tanaka S (1995b) Does silicon exist in association with organic compounds in rice plant? Jpn J Soil Sci Plant Nutr 11:111–117

Ishiguro K (2001) Review of research in Japan on the roles of silicon in conferring resistance against rice blast. In: Datnoff LE, Snyder GH, Korndörfer GH (eds) Silicon in agriculture, studies in Plant Science 8. Elsevier Science, Amsterdam, pp 277–291

Ito S, Hayashi H (1931) On the relation of silica supply to rice blast. J Sapporo Soc Agric Sci 103:460–461

Ito S, Sakamoto M (1939) Studies on rice blast. Hokkaido University, Hokkaido

Jiang D (1993) The role of silicon in barley resistance to powdery mildew (*Erysiphe graminis* f. sp. *hordei*). PhD thesis, University of Minnesota, St. Paul

Kauss H, Seehaus K, Franke R, Gilbert S et al (2003) Silica deposition by a strongly cationic proline-rich protein from systemically resistant cucumber plants. Plant J 33:87–95. doi:10.1046/j.1365-313X.2003.01606.x

Kawamura E, Ono K (1948) Study on the relation between the pre-infection behavior of rice blast fungus, *Piricularia oryzae*, and water droplets on rice plant leaves. J Natl Agr Expt Sta 4:1–12

Kim SG, Kim KW, Park EW et al (2002) Silicon-induced cell wall fortification of rice leaves: a possible cellular mechanism of enhanced host resistance to blast. Phytopathology 92:1095–1103. doi:10.1094/PHYTO.2002.92.10.1095

Koga H, Zeyen RJ, Bushnell WR et al (1988) Hypersensitive cell death, autofluorescence, and insoluble silicon accumulation in barley leaf epidermal cells under attack *by Erysiphe graminis* f. sp. *hordei*. Physiol Mol Plant Pathol 32:395–409

Lamb C, Dixon RA (1997) The oxidative burst in plant disease resistance. Annu Rev Plant Biol 48:251–275. doi:10.1146/annurev.arplant.48.1.251

Less H, Angelovici R, Tzin V et al (2011) Coordinated gene networks regulating Arabidopsis plant metabolism in response to various stresses and nutritional cues. Plant Cell 23:1264–1271. doi:10.1105/tpc.110.082867

Li A, Heath MC (1990) Effect of plant growth regulators on the interactions between bean plants and rust fungi non-pathogenic on beans. Physiol Mol Plant Pathol 37:245–254

Li W, Bi Y, Ge Y et al (2012) Effects of postharvest sodium silicate treatment on pink rot disease and oxidative stress-antioxidative system in muskmelon fruit. Eur Food Res Technol 234:137–145. doi:10.1007/s00217-011-1611-9

Liang YC, Sun WC, Si J, Römheld V (2005) Effects of foliar-and root-applied silicon on the enhancement of induced resistance to powdery mildew in *Cucumis sativus*. Plant Pathol 54:678–685. doi:10.1111/j.1365-3059.2005.01246.x

Liu M, Cai K, Chen Y et al (2014) Proteomic analysis of silicon-mediated resistance to *Magnaporthe oryzae* in rice (*Oryza sativa* L.). Eur J Plant Pathol 139:579–592. doi:10.1007/s10658-014-0414-9

Ma JF, Yamaji N (2006) Silicon uptake and accumulation in higher plants. Trends Plant Sci 11:392–397. doi:10.1016/j.tplants.2006.06.007

Maekawa K, Watanabe K, Kanto T et al (2002) Accumulation of silicon around penetration sites of *Magnaporthe grisea* and silicon-dependent promotion of superoxide generation after inoculation on rice leaf. In: Matoh T (ed) Second silicon in agriculture conference. Press-Net, Kyoto, pp 34–38

Menzies JG, Ehret DL, Glass ADM, Samuels AL et al (1991) The influence of silicon on cytological interactions between *Sphaerotheca fuliginea* and *Cucumis sativus*. Physiol Mol Plant Pathol 39:403–414. doi:10.1016/0885-5765(91)90007-5

Mohaghegh P, Khoshgoftarmanesh AH, Shirvani M et al (2011) Effect of silicon nutrition on oxidative stress induced by *Phytophthora melonis* infection in cucumber. Plant Dis 95:455–460. doi:10.1094/PDIS-05-10-0379

Motomura H, Fujii T, Suzuki M (2004) Silica deposition in relation to ageing of leaf tissues in *Sasa veitchii* (Carrière) Rehder (Poaceae: Bambusoideae). Ann Bot 93:235–248. doi:10.1093/aob/mch034

Nascimento KJT, Debona D, França SKS et al (2014) Soybean resistance to *Cercospora sojina* infection is reduced by silicon. Phytopathology 104:1183–1191. doi:10.1094/PHYTO-02-14-0047-R

Pereira SC, Rodrigues FA, Carré-Missio V et al (2009a) Aplicação foliar de silício na resistência da soja à ferrugem e na atividade de enzimas de defesa. Trop Plant Pathol 34:164–170. doi:10.1590/S1982-56762009000300005

Pereira SC, Rodrigues FA, Carré-Missio V et al (2009b) Efeito da aplicação foliar de silício na resistência à ferrugem e na potencialização da atividade de enzimas de defesa em cafeeiro. Trop Plant Pathol 34:223–230. doi:10.1590/S1982-56762009000400004

Perumalla CJ, Heath MC (1991) The effect of inhibitors of various cellular processes on the wall modifications induced in bean leaves by the cowpea rust fungus. Physiol Mol Plant Pathol 38:293–300. doi:10.1016/S0885-5765(05)80120-1

Petit AN, Vaillant N, Boulay M et al (2006) Alteration of photosynthesis in grapevines affected by esca. Phytopathology 96:1060–1066. doi:10.1094/PHYTO-96-1060

Polanco LR, Rodrigues FA, Nascimento KJT et al (2014) Photosynthetic gas exchange and antioxidative system in common bean plants infected by *Colletotrichum lindemuthianum* and supplied with silicon. Trop Plant Pathol 39:35–42. doi:10.1590/S1982-56762014000100005

Pozza AAA, Alvez E, Pozza EA et al (2004) Efeito do silício no controle da cercosporiose em três variedades de cafeeiro. Fitopatol Bras 29:185–188. doi:10.1590/S0100-41582004000200010

Rahman A, Wallis C, Uddin W (2015) Silicon induced systemic defense responses in perennial ryegrass against infection by *Magnaporthe oryzae*. Phytopathology 105:748–757. doi:10.1094/PHYTO-12-14-0378-R

Rémus-Borel W, Menzies JG, Bélanger RR (2005) Silicon induces antifungal compounds in powdery mildew-infected wheat. Physiol Mol Plant Pathol 66:108–115. doi:10.1094/PDIS-04-11-0274

Rémus-Borel W, Menzies JG, Bélanger RR (2009) Aconitate and methyl aconitate are modulated by silicon in powdery mildew-infected wheat plants. J Plant Physiol 166:1413–1422. doi:10.1016/j.jplph.2009.02.011

Resende RS, Rodrigues FA, Cavatte PC et al (2012) Leaf gas exchange and oxidative stress in sorghum plants supplied with silicon and infected by *Colletotrichum sublineolum*. Phytopathology 102:892–898. doi:10.1094/PHYTO-01-12-0014-R

Resende RS, Rodrigues FA, Gomes RJ, Nascimento KJT et al (2013) Microscopic and biochemical aspects of sorghum resistance to anthracnose mediated by silicon. Ann Appl Biol 163:114–123. doi:10.1111/aab.12040

Rezende DC, Rodrigues FA, Carré-Missio V, Schurt DA, Kawamura K, Korndörfer GH (2009) Effect of root and foliar applications of silicon on brown spot development in rice. Australas Plant Pathol 38:67–73. doi:10.1071/ap08080

Rios JA, Rodrigues FA, Debona D et al (2014) Photosynthetic gas exchange in leaves of wheat plants supplied with silicon and infected with *Pyricularia oryzae*. Acta Physiol Plant 36:371–379. doi:10.1007/s11738-013-1418-3

Rodrigues FA, Benhamou N, Datnoff LE, Jones JB, Bélanger RR (2003) Ultrastructural and cytochemical aspects of silicon-mediated rice blast resistance. Phytopathology 93:535–546

Rodrigues FA, McNally DJ, Datnoff LE et al (2004) Silicon enhances the accumulation of diterpenoid phytoalexins in rice: a potential mechanism for blast resistance. Phytopathology 94:177–183

Rodrigues FA, Jurick WM, Datnoff LE et al (2005) Silicon influences cytological and molecular events in compatible and incompatible rice-*Magnaporthe grisea* interactions. Physiol Mol Plant Pathol 66:144–159

Rodrigues FA, Polanco LR, Duarte HSS et al (2015) Photosynthetic gas exchange in common bean submitted to foliar sprays of potassium silicate, sodium molybdate and fungicide and infected with *Colletotrichum lindemuthianum*. J Phytopathol 163:554–559. doi:10.1111/jph.12353

Rojas CM, Senthil-Kumar M, Tzin V, Kirankumar S et al (2014) Regulation of primary plant metabolism during plant-pathogen interactions and its contribution to plant defense. Front Plant Sci 5:17. doi:10.3389/fpls.2014.00017

Samuels AL, Glass ADM, Ehret DL et al (1991) Mobility and deposition of silicon in cucumber plants. Plant Cell Environ 41:485–492. doi:10.1111/j.1365-3040.1991.tb01518.x

Scholes JD, Rolfe SA (1996) Photosynthesis in localised regions of oat leaves infected with crown rust (*Puccinia coronata*): quantitative imaging of chlorophyll fluorescence. Planta 199:573–582

Schurt DA, Rodrigues FA, Reis RD et al (2012) Resistência física de bainhas de plantas de arroz supridas com silício e infectadas por *Rhizoctonia solani*. Trop Plant Pathol 37:281–285. doi:10.1590/S1982-56762012000400008

Schurt DA, Rodrigues FA, Carré-Missio V et al (2013a) Silício alterando compostos derivados da pirólise de bainhas foliares de plantas de arroz infectadas por *Rhizoctonia solani*. Bragantia 72:52–60

Schurt DA, Rodrigues FA, Colodette JL et al (2013b) Efeito do silício nas concentrações de lignina e de açúcares em bainhas de folhas de arroz infectadas por *Rhizoctonia solani*. Bragantia 72:360–366

Schurt DA, Cruz MF, Nascimento KJT et al (2014) Silicon potentiates the activities of defense enzymes in the leaf sheaths of rice plants infected by *Rhizoctonia solani*. Trop Plant Pathol 39:457–463. doi:10.1590/S1982-56762014000600007

Schurt DA, Reis RD, Araujo L, Carré-Missio V, Rodrigues FA (2015) Análise microscópica da resistência do arroz à queima das bainhas mediada pelo silício. Bragantia 74:93–101, doi: http://dx.doi.org/10.1590/1678-4499.0355

Seebold KW, Kucharek TA, Datnoff LE et al (2001) The influence of silicon on components of resistance to blast in susceptible, partially resistant, and resistant cultivars of rice. Phytopathology 91:63–69. doi:10.1094/PHYTO.2001.91.1.63

Shetty R, Fretté X, Jensen B et al (2011) Silicon-induced changes in antifungal phenolic acids, flavonoids, and key phenylpropanoid pathway genes during the interaction between miniature roses and the biotrophic pathogen *Podosphaera pannosa*. Plant Physiol 157:2194–2205. doi:10.1104/pp.111.185215

Shetty R, Jensen B, Shetty NP et al (2012) Silicon induced resistance against powdery mildew of roses caused by *Podospahera pannosa*. Plant Pathol 61:120–131. doi:10.1111/j.1365-3059.2011.02493.x

Shtienberg D (1992) Effects of foliar diseases on gas exchange processes: a comparative study. Phytopathology 82:760–765. doi:10.1094/Phyto-82-760

Silva RV, Oliveira RDL, Nascimento KJT, Rodrigues FA et al (2010) Biochemical responses of coffee resistance against *Meloidogyne exigua* mediated by silicon. Plant Pathol 59:586–593. doi:10.1111/j.1365-3059.2009.02228.x

Silva MRJ, Pereira SC, Rodrigues FA, Zanão Júnior KA et al (2012) Silicon and manganese on the activity of enzymes involved in rice resistance against brown spot. Trop Plant Pathol 37:339–345. doi:10.1590/S1982-56762012000500006

Sousa RS, Rodrigues FA, Schurt DA, Souza NFA, Cruz MFA (2013) Cytological aspects of the infection process of *Pyricularia oryzae* on leaves of wheat plants supplied with silicon. Trop Plant Pathol 38:472–477. doi:10.1590/S1982-56762013000600002

Sun W, Zhang J, Fan Q et al (2010) Silicon-enhanced resistance to rice blast is attributed to silicon-mediated defence resistance and its role as physical barrier. Eur J Plant Pathol 128:39–49. doi:10.1007/s10658-010-9625-x

Suzuki H (1940) On the relationship between rice susceptibility and penetration into host plants. Ngyo Oyobi Engei 10:1999–2010

Swarbrick PJ, Schulze-Lefert P, Scholes JD (2006) Metabolic consequences of susceptibility and resistance (race-specific and broad-spectrum) in barley leaves challenged with powdery mildew. Plant Cell Environ 29:1061–1076

Tatagiba SD, Rodrigues FA, Filippi MCC et al (2014) Physiological responses of rice plants supplied with silicon to *Monographella albescens* infection. J Phytopathol 162:596–606. doi:10.1111/jph.12231

Torres MA, Jones JDG, Dangl JL (2006) Reactive oxygen species signaling in response to pathogens. Plant Physiol 141:373–378. doi:10.1104/pp.106.079467

Van Bockhaven J, De Vleesschauwer D, Höfte M (2013) Towards establishing broad-spectrum disease resistance in plants: silicon leads the way. J Exp Bot 64:1281–1293. doi:10.1093/jxb/ers329

Van Bockhaven J, Spíchal L, Novák O et al (2015) Silicon induces resistance to the brown spot fungus *Cochliobolus miyabeanus* by preventing the pathogen from hijacking the rice ethylene pathway. New Phytol 206:761–773

Van Hulten M, Pelser M, Van Loon LC et al (2006) Costs and benefits of priming for defense in Arabidopsis. Proc Natl Acad Sci U S A 103:5602–5607

Vivancos J, Labbé C, Menzies JG, Bélanger RR et al (2015) Silicon-mediated resistance of Arabidopsis against powdery mildew involves mechanisms other than the salicylic acid (SA)-dependent defence pathway. Mol Plant Pathol 16:572–582. doi:10.1111/mpp.12213

Volk RJ, Kahn RP, Weintraub RL (1958) Silicon content of the rice plant as a factor influencing its resistance to infection by the blast fungus, *Piricularia oryzae*. Phytopathology 48:179–184. doi:10.1590/S0100-41582005000500001

Yoshida S, Ohnishi Y, Kitagishi K (1962) Chemical forms, mobility, and deposition of silicon in the rice plant. Jpn JSoil Sci Plant Nutr 8:107–111

Zeyen RJ, Ahlstrand GG, Carver TLW (1993) X-ray microanalysis of frozen-hydrated, freeze-dried, and critical point dried leaf specimens: determination of soluble and insoluble chemical elements at *Erysiphe graminis* epidermal cell papilla sites in barley isolines containing *Ml-o* and *ml-o* alleles. Can J Bot 71:284–296

Zhao D, Glynn NC, Glaz B et al (2011) Orange rust effects on leaf photosynthesis and related characters of sugarcane. Plant Dis 95:640–647. doi:10.1094/PDIS-10-10-0762

Chapter 6
Highlights and Prospects for Using Silicon in the Future

Lawrence E. Datnoff and Fabrício A. Rodrigues

Abstract As a result of silicon research from the 1980s until today, a number of facts can be stated about the role this element plays in plant disease suppression. These include the following: for any plant disease, a minimum silicon concentration is needed to suppress that disease; once that level has been obtained, plant disease suppression increases proportionally as the silicon concentration (insoluble or soluble) increases in plant tissues; the silicon supply to a plant must be continuous or the disease-suppressing effects will be reduced or non-existent; silicon can influence many components of host resistance; silicon may augment susceptible and partial resistance almost at the same level as complete genetic resistance; only when applying silicon to the roots will this element mediate plant defenses at both the physiological and molecular level; and silicon may suppress plant diseases as effectively as fungicides. In spite of the recent advances linking silicon to host resistance via the -"omics", namely, genomics, proteomics and metabolomics, the exact mechanism(s) by which this element modulates plant physiology through an increase in host resistance requires further investigation. Silicon undoubtedly deserves more attention by scientists and agriculturalists, but its recognition is limited by current perspectives on whether agricultural soils are truly low in this element, whether the plant in question will accumulate silicon and whether silicon is to be viewed as a fertilizer, biostimulant or plant protectant. Nevertheless, as researchers and growers become more aware of silicon and its potential, it is likely that this often overlooked, quasi-essential element will be recognized as a viable means of enhancing plant health and performance.

L.E. Datnoff (✉)
Department of Plant Pathology and Crop Physiology, Louisiana State University Agricultural Center, 302 Life Sciences, Baton Rouge, LA 70803, USA
e-mail: LDatnoff@agcenter.lsu.edu

F.A. Rodrigues
Departamento de Fitopatologia, Universidade Federal de Viçosa, Viçosa, Minas Gerais Cep 36570-900, Brazil
e-mail: fabricio.plantpathologist@gmail.com; fabricio@ufv.br

F.A. Rodrigues, L.E. Datnoff (eds.), *Silicon and Plant Diseases*,
DOI 10.1007/978-3-319-22930-0_6

Introduction

Silicon may dramatically decrease the intensity of a number of seed, soil-borne and foliar diseases caused by a plethora of plant pathogens that attack a number of important agronomic and horticultural crops. On the basis of research conducted to date, a better understanding about silicon in relation to plant disease suppression is known and includes the following:

- for any plant disease, a minimum silicon concentration is needed to suppress that disease (Dallagnol et al. 2009);
- once that level has been obtained, plant disease suppression increases proportionally as the silicon concentration (insoluble or soluble) increases in plant tissues (Datnoff et al. 1991; Sun et al. 2010);
- the silicon supply to the plant must be continuous or the disease-suppressing effects will be reduced or non-existent (Heine et al. 2007; Samuels et al. 1991);
- silicon will influence many components of host resistance that include the incubation period, latent period, lesion number and lesion size (Brecht et al. 2007; Resende et al. 2013; Rodrigues et al. 2001; Seebold et al. 2001);
- silicon may augment susceptible and partial resistance almost at the same level as complete genetic resistance (Resende et al. 2013; Rodrigues et al. 2001; Seebold et al. 2000);
- only when applying silicon to the roots will this element change plant responses to pathogen infections at both the physiological and molecular levels, implying an active role for silicon in one or more plant defense signaling pathways (Brunings et al. 2009; Rodrigues et al. 2004, 2005); and
- silicon may suppress plant diseases as effectively as fungicides (Brecht et al. 2004; Resende et al. 2013; Seebold et al. 2004).

What Information Do We Still Need to Improve the Deployment of Silicon for Managing Plant Diseases?

As mentioned in the other chapters of this book, silicon cannot be applied to foliage to potentiate the activation of host defense mechanisms in the same way as root applications because there are currently no known transporter genes to move this element through the cuticle followed by basipetal movement into the roots. Perhaps nanotechnology could provide a solution to move silicon into the plant through its foliage. Nanoparticles (NP) possess unique chemical and physical properties, and research has already demonstrated the enhanced availability and transport of copper (Cu) nanoparticles through a foliar route (Elmer et al. 2014). In this study, the authors reported that a higher Cu level was detected in the roots of the NP Cu foliar-treated plants in comparison with the bulked equivalent Cu treatment or the control. As a consequence, the progress of Fusarium wilt in tomato plants was more effectively suppressed. There is experimental evidence to suggest that this suppression might be stimulated with silicon nanoparticles as shown with Cu (Suriyaprabha et al. 2012). Suriyaprabha and colleagues (2012) demonstrated that silicon accumulation in maize leaves was increased as the NP concentration increased, and it ranged from 0.57 % to 0.82 %. Along with this increase in silicon accumulation, maize growth was found to be positively influenced. Undoubtedly, further research is warranted to determine if this NP foliar approach would suppress plant diseases as effectively as applying silicon to the roots.

In spite of recent advances linking silicon to host resistance through the -omics, namely genomics, proteomics and metabolomics, the exact mechanism(s) by which this element modulates plant physiology through increased host resistance requires further investigation (Bockhaven et al. 2012). The information generated so far has provided novel insights into the potential for silicon to interact with multiple pathways of the plant's primary metabolism to help plants cope better with infections of both soil-borne and foliar pathogens. Considering the current plant nutriomics scenario, it will be necessary to determine the involvement of silicon in plant-signaling pathways, which will lead to the potentiation of host defense mechanisms and also make it feasible to modify key regulator(s) to enhance silicon uptake. In the near future, it will be possible to elucidate the real functions of silicon at the molecular, cellular, organ and even whole plant levels.

Why Is Silicon Still Not Used Routinely for Managing Plant Diseases Under Greenhouse and Field Conditions?

What is holding producers and growers back from using silicon? One possible answer is its lack of recognition. Silicon is not recognized as being a limiting nutrient in the soil because it is known to be the second-most common element in the earth's crust after oxygen. Even so, a number of soil orders have been shown to be low or limiting in this element such as sandy Entisols, Histosols, acidic Inceptisols, Oxisols and Ultisols. This perspective is related, in part, to the ways in which essential plant nutrients are defined as based on three criteria developed by Arnon and Stout (Epstein and Bloom 2005) that must be met, namely, "1) a deficiency of the element makes it impossible for the plant to complete its life cycle; 2) the deficiency is specific for the element in question and 3) the deficiency is directly involved in the nutrition of the plant as for example as a constituent of an essential metabolite or required for the action of an enzyme system". Epstein and Bloom (2005) have argued that there are difficulties with this definition. For the first criterion, a plant can be quite severely deficient in an essential nutrient element and still be able to complete its life cycle. The second criterion implies that the element for which it is substituted is *"deficient"*, and thus this criterion merely reiterates the first requirement. Finally, the third criterion, in which the element is directly involved in plant nutrition, discounts the correction of unfavorable environmental conditions. This inherent disregard has not always occurred during the discovery of other essential elements (Epstein and Bloom 2005). When boron was discovered to be essential, no one had any evidence that "the element is directly involved in the nutrition of the plant". Based on these deficiencies in the Arnon-Stout definition, Epstein and Bloom (2005) have proposed the following criteria: "An element is essential if it fulfills either one or both of two criteria: (1) The element is part of a molecule that is an intrinsic component of the structure or metabolism of a plant; and (2) the plant can be so severely deprived of the element that it exhibits abnormalities in its growth, development or reproduction – that is, its "performance" – in comparison with plants not so deprived". Because the essentiality of silicon for diatoms and *Equisetum arvense*, or 'scouring rushes', is well established but has not been categorically demonstrated for other silicon-accumulating plant species, Epstein and Bloom (2005) have proposed that silicon is a 'quasi-essential' element. Therefore, this perception of silicon must change if this element is to be used in agriculture, especially for suppressing plant diseases.

Another reason why silicon is not used routinely for plant disease management is that it is not clear whether a plant does or does not accumulate this element. The criteria proposed to distinguish accumulating from non-accumulating plants were that accumulators have a silicon concentration greater than 1 dag/kg and a silicon/calcium ratio greater than 1, and excluders have a silicon concentration below 0.5 dag/kg and a silicon/calcium ratio less than 0.5 (Takahasi et al. 1990). Plants that do not meet either one of these criteria would be known as intermediates. When determining whether a plant accumulates silicon, past studies have focused primarily on measuring silicon in the foliage and did not routinely measure it in other plant organs. Some plant species such as Chinese cabbage, clover, coffee, crimson, green onions, peppers, radishes and tomatoes have been shown to concentrate more silicon in their roots than in their shoots (Lewin and Reimann 1969; Carré-Missio et al. 2009; French-Monar et al. 2010; Huang et al. 2011). Therefore, based on the above criteria, these plants would be considered to be excluders or rejecters. Currently, all plants that are grown in soil are known to contain some silicon in their tissues, and 44 angiosperm clades (representing over 100 orders or families) have been reported to date to contain silicon (Hodson et al. 2005). This study by Hodson and his colleagues was, however, based solely on leaf tissue that was sampled from a number of plant species to determine whether they were silicon accumulators or not on the basis of the above criteria. Even so, by knowing which plant tissues better accumulate this element, this determination might help in deploying silicon for managing plant diseases. For example, tomatoes accumulate more silicon in the roots than in the foliage. Therefore, if Fusarium crown and root rot is a potential problem in the field and the soils are known to be low or limiting in silicon, then fertilizing with this element could help to reduce the intensity of this disease (Huang et al. 2011).

Another problem has to do with the classification of silicon as a fertilizer as opposed to a plant protectant such as a fungicide. Even so, some individuals in the horticultural industry are beginning to suggest that silicon should be classified as a biostimulant. The problem with describing it as a biostimulant as well as using the fertilizer label allows these classifications to be used only for a silicon product that will be marketed for use in combating abiotic stresses such as heavy metal toxicities or temperature extremes, and not for plant disease control. If it is marketed for plant disease control, at least in the USA, the product must be registered as a plant protectant (pesticide). Many countries still do not have a way to classify silicon at all. To the best of our current knowledge, Japan and Korea recognized the importance of silicon in rice back in the 1950s, and they classified this element as being agronomically essential. In 2004, Brazil was the third country to formally recognize silicon. The Brazilian Ministry of Agriculture, which regulates the commercial production of fertilizers, ruled that silicon is a beneficial micronutrient. In 2013, the American Association for Plant Food Control Officials (AAPFCO), the regulatory body that governs the labeling of fertilizers in the USA, recognized silicon as a beneficial substance that can now be sold as a fertilizer across the USA. In 2007, potassium silicate (Sil-MATRIX^{R}, PQ Corporation) was registered by the EPA in the US and certified by OMRI as an organic pesticide for the control of powdery mildew and the control of mites and aphids on high-value crops such as strawberries, wine grapes and others. Without a doubt, silicon is a cross-over element that acts as both a fertilizer and a plant protectant because it affects both abiotic and biotic stresses. Perhaps a new category should be created so that silicon may achieve universal acceptance by regulatory agencies worldwide.

Conclusions

The idea that silicon plays an important role in the mineral nutrition of plants is not in doubt, nor is its ability to enhance plant development and efficiently decrease the intensity of plant diseases (Datnoff 2005). More evidence is now accumulating to show that the effects of this element in terms of disease suppression strongly impact a large number of monocot and dicot species, and our understanding of how it functions in the plant is greatly improving. Effective, practical means of application, affordable sources of silicon and methods for identifying conditions under which silicon fertilization will be beneficial are still needed for use in disease management. However, research on the use of silicon for plant disease suppression under field conditions is still in its infancy. For example, no soil tests for gauging the amounts of plant-available silicon have been calibrated for many agronomic or horticulture crops. Furthermore, most analytical laboratories do not routinely assay plant tissue for silicon. In fact, the current standard tissue digestion procedures used in most laboratories would render silicon insoluble, making an analysis of the digested tissue meaningless. Thus, the two analytical tools most often used for determining the need for fertilization with plant nutrients are not widely available for silicon. Although plant disease suppression by silicon applications has been documented in a number of controlled experiments, particularly in the greenhouse, only a few large-scale field effects have been observed to date for rice, sorghum and sugarcane. The conditions under which disease suppression with silicon fertilization will occur are not well known for a number of other agronomic and horticultural crops. Nevertheless, as the need for environmentally friendly strategies for the management of biotic stress increases, silicon could provide a valuable tool. The use of silicon for controlling plant diseases while improving plant performance would be well-suited for inclusion in integrated disease management strategies and would permit potential reductions in fungicide use as well as the enhancement of host plant resistance. As researchers and growers become more aware of silicon and its potential, it is likely that this often overlooked, quasi-essential element will be recognized as a viable means of enhancing plant health and performance.

References

Bockhaven JV, Vleeschauwer DD, Höfte M (2012) Towards establishing broad-spectrum disease resistance in plants: silicon leads the way. J Exp Bot 64:1281–1293

Brecht MO, Datnoff LE, Kucharek TA, Nagata RT (2004) Influence of silicon and chlorothalonil on plant growth and suppression of gray leaf spot of St. Augustinegrass. Plant Dis 88:338–344

Brecht MO, Datnoff LE, Kucharek TA, Nagata RT (2007) Effect of silicon on components of resistance to gray leaf spot in St. Augustinegrass. J Plant Nutr 30:1005–1021

Brunings AM, Datnoff LE, Ma JF, Mitani N, Nagamura Y, Rathinasabapathi B, Kirst K (2009) Differential gene expression of rice in responses to silicon and the rice blast fungus *Magnaporthe oryzae*. Ann App Biol 155:161–170

Carré-Missio V, Rodrigues FA, Schurt DA, Pereira SC, Oliveira MGA, Zambolim L (2009) Ineficiência do silício no controle da ferrugem do cafeeiro em solução nutritiva. Trop Plant Pathol 34:416–421. doi:10.1590/S1982-56762009000600009

Dallagnol LJ, Rodrigues FA, Mielli MVB, Ma JF, Datnoff LE (2009) Defective active silicon uptake affects some components of rice resistance to brown spot. Phytopathology 99:116–121

Datnoff LE (2005) Silicon in the life and performance of turfgrass. Online. Applied Turfgrass Science. doi:10.1094/ATS-2005-0914-01-RV

Datnoff LE, Raid RN, Snyder GH, Jones DB (1991) The effect of calcium silicate on blast and brown spot intensities and yields of rice. Plant Dis 74:729–732

Elmer W, Servin A, White J (2014) Use nanoparticles of Cu, Mn, and Zn to suppress soilborne diseases of eggplants and tomatoes. In: 16th world fertilizer congress of CIEC, Technological innovation for a sustainable tropical agriculture, 20–24 Oct, Rio de Janeiro, pp 87–89

Epstein E, Bloom AJ (2005) Mineral nutrition of plants: principles and perspectives, 2nd edn. Sinauer Associates, Sunderland

French-Monar R, Rodrigues FA, Korndörfer GH, Datnoff LE (2010) Silicon suppresses Phytophthora blight development on bell pepper. J Phytopathol 158:554–560

Heine G, Tikum G, Horst WJ (2007) The effect of silicon on the infection by and spread of *Pythium aphanidermatum* in single roots of tomato and bitter gourd. J Exp Bot 58:569–577

Hodson MJ, White PJ, Mead A (2005) Phylogenetic variation in the silicon composition of plants. Ann Bot 96:1027–1046

Huang CH, Roberts PD, Datnoff LE (2011) Silicon suppresses Fusarium crown and root rot of tomato. J Phytopathol 159:546–554

Lewin JL, Reimann EF (1969) Silicon and plant growth. Annu Rev Plant Physiol 20:289–304

Resende RS, Rodrigues FA, Costa RV, Silva DD (2013) Silicon and fungicide effects on anthracnose in moderately resistant and susceptible sorghum lines. J Phytopathol 161:11–17

Rodrigues FA, Datnoff LE, Korndörfer GH, Seebold KW, Rush MC (2001) Effect of silicon and host resistance on sheath blight development in rice. Plant Dis 85:827–832

Rodrigues FA, McNally DJ, Datnoff LE, Jones JB, Labbe C, Benhamou N, Menzies JG, Bélanger RR (2004) Silicon enhances the accumulation of diterpenoid phytoalexins in rice: a biochemical mechanism for blast resistance. Phytopathology 94:177–183
Rodrigues FA, Jurick WM, Datnoff LE, Jones JB, Rollins JA (2005) Cytological and molecular aspects of silicon-mediated resistance in rice against *Magnaporthe grisea*. Physiol Mol Plant Pathol 66:144–159
Samuels AI, Glass ADM, Ehret DL, Menzies JG (1991) Mobility and deposition of silicon in cucumber plants. Plant Cell Environ 14:485–492
Seebold KW, Datnoff LE, Correa-Victoria FJ, Kucharek TA, Snyder GH (2000) Effect of silicon rate and host resistance on blast, scald, and yield of upland rice. Plant Dis 84:871–876
Seebold KW, Kucharek TA, Datnoff LE, Correa-Victoria FJ, Marchetti MA (2001) The influence of silicon on components of resistance to blast in susceptible, partially resistant and resistant cultivars of rice. Phytopathology 91:63–69
Seebold KW, Datnoff LE, Correa-Victoria FJ, Kucharek TA, Snyder GH (2004) Effects of silicon and fungicides at full and reduced rates on the control of leaf and neck blast in upland rice. Plant Dis 88:253–258
Sun W, Zhang J, Fan Q, Xue G, Li Z, Liang Y (2010) Silicon-enhanced resistance to rice blast is attributed to silicon-mediated defense resistance and its role as physical barrier. Eur J Plant Pathol 128:39–49. doi:10.1007/s10658-010-9625-x
Suriyaprabha, R, Karunakaran G, Yuvakkumar R, Prabu P, Rajendran V, Kannan N (2012) Growth and physiological responses of maize (*Zea mays* L.) to porous silica nanoparticles in soil. J Nanopart Res 14:1294. doi:10.1007/s11051-012-1294-6
Takahasi E, Ma JF, Miyake Y (1990) The possibility of silicon as an essential element for higher plants. Commen Agric Food Chem 2:99–122

Index

F.A. Rodrigues, L.E. Datnoff (eds.), *Silicon and Plant Diseases*,
DOI 10.1007/978-3-319-22930-0

The manufacturer's authorised representative in the EU is Springer Nature Customer Service Centre GmbH, Europaplatz 3, 69115 Heidelberg, Germany. If you have any concerns regarding our products, please contact ProductSafety@springernature.com

Printed and bound by CPI Group (UK) Ltd, Croydon, CR0 4YY
05/10/2026
02228157-0003